John Ball

Notes of a Naturalist in South America

John Ball

Notes of a Naturalist in South America

ISBN/EAN: 9783337313517

Printed in Europe, USA, Canada, Australia, Japan

Cover: Foto ©berggeist007 / pixelio.de

More available books at **www.hansebooks.com**

NOTES OF A NATURALIST
IN SOUTH AMERICA

NOTES OF A NATURALIST

IN SOUTH AMERICA

BY

JOHN BALL, F.R.S., M.R.I.A., ETC.

LONDON
KEGAN PAUL, TRENCH & CO., 1, PATERNOSTER SQUARE
1887

TO

L. M.,

WHOSE SUGGESTIONS LED TO ITS TAKING SHAPE,

I DEDICATE THIS LITTLE BOOK.

PREFACE.

A TOUR round the South American continent, which was completed in so short a time as five months, may not appear to deserve any special record; yet I am led to hope that this little book may serve to induce others to visit a region so abounding in sources of enjoyment and interest. There is no part of the world where, in the same short space of time, a traveller can view so many varied and impressive aspects of nature; while he whose attention is mainly given to the progress and development of the social condition of mankind will find in the condition of the numerous states of the continent, and the manners and habits of the many different races that inhabit it, abundant material to engage his attention and excite his interest.

Although, as the title implies, the aim of my journey was mainly directed to the new aspects of nature, organic and inorganic, which South America superabundantly presents to the stranger, I have not thought it without interest to give in these pages the impressions as to the social and political condition of the different regions which I visited, suggested to an unprejudiced visitor by the daily incidents of a traveller's life.

Those who may be tempted to undertake a tour in South America will find that by a judicious choice of route, according to the season selected for travelling, they may visit all the accessible parts of the continent with perfect ease, and with no more risk of injury to health, or of bodily discomfort, than they incur in a summer excursion in Europe. The chief precaution to be observed is to make the visit to Brazil fall in the cool and dry season, extending from mid-May to September. It may also be well to mention that, while the cost of passage and expenses on board, for a journey of about 18,400 miles by sea, somewhat exceeded £170, my expenses during about ten weeks on land, without any attempt at economy, did not exceed £100.

The reader may regard as superfluous the rather frequent references to the meteorology of the various parts of the continent which I was able to visit. But, if he will consider the importance of the two main elements—temperature and moisture—in regulating the development of organic life in past epochs, and the influence which they now exercise on the character of the human population, he will admit that a student of nature could not fail to make them the objects of frequent attention, the more especially as many erroneous impressions as to the climate of various parts of South America are still current, even among men of science.

I make no pretension to add anything of importance to our store of positive knowledge respecting the region described in this volume; I shall be content if it should be found that I have suggested trains of thought that may lead others to valuable results. I venture, indeed, to believe that the argument adduced in the sixth chapter, as to the great extent and importance of the ancient mountains of Brazil, approaches near to demonstration, and that the recognition of its validity will be found to throw fresh light on the history of organic life in that region of the globe.

In the Appendices to this volume two subjects of a somewhat technical character, not likely to interest the general reader, are separately discussed. With regard to both of them, my aim has been to show that the opinions now current amongst men of science do not rest upon adequate evidence, and that we need further knowledge of the phenomena, discoverable by observation, before we can safely arrive at positive conclusions.

In deference to the prejudices of English readers, which are unfortunately shared by many scientific writers, the ordinary British standards of measure and weight have been followed throughout the text, as well as the antiquated custom of denoting temperature by the scale of Fahrenheit's thermometer. With regard to the metrical system of measures and weights, I am fully aware of its imperfections, and if the question were now raised for the first time I should advocate the adoption of some considerable modifications. But seeing that no other uniform system is in existence, and that the metrical system has been adopted by nearly all civilized nations, I cannot but regret that my countrymen should retain what is practically a barrier to the free interchange of thought with the rest of the world. The defects of the metrical system are mainly those of our decimal system of numeration, which owes its existence to the fact that the human hand possesses five fingers. If in some future stage of development our race should acquire a sixth finger to each hand, it may then also acquire a more convenient system of numeration, to which the scale of measures would naturally be adapted. In the mean time the advantages of a uniform system far outweigh its attendant defects.

The adherence to the Fahrenheit scale for the thermometer is even less defensible. It belongs to a primitive epoch of science, when a knowledge of the facts of physics was in a rudimentary stage, and its survival at the present day is a matter of marvel to the student of progress.

I should not conclude these prefatory words without expressing my obligations to many scientific friends whom I have from time to time consulted with advantage; and I must especially record my obligation to Mr. Robert Scott, F.R.S., who has on many occasions been my guide to the valuable materials available in the library of the Meteorological Office.

CONTENTS.

CHAPTER I.

Voyage across the Atlantic—Barbadoes—Jamaica—Isthmus of Panama—Buenaventura, tropical forest—Guayaquil and the river Guayas—Payta—The rainless zone of Peru—Voyage to Callao 1

CHAPTER II.

Arrival at Callao—Quarantine—The war between Chili and Peru—Aspect of Lima—General Lynch—Andean railway to Chicla—Valley of the Rimac—Puente Infernillo—Chicla—Mountain-sickness—Flora of the Temperate zone of the Andes—Excursion to the higher region—Climate of the Cordillera—Remarks on the Andean flora—Return to Lima—Visit to a sugar-plantation—Condition of Peru—Prospect of anarchy 56

CHAPTER III.

Voyage from Callao to Valparaiso—Arica—Tocopilla—Scenery of the moon—Caldera—Aspect of North Chili—British Pacific squadron—Coquimbo—Arrival at Valparaiso—Climate and vegetation of Central Chili—Railway journey to Santiago—Aspect of the city—Grand position of Santiago—Dr. Philippi—Excursion to Cerro St. Cristobal—Don B. Vicuña Mackenna—Remarkable trees—Excursion to the baths of Cauquenes—The first rains—Captive condors—Return to Santiago—Glorious sunset 118

CHAPTER IV.

PAGE

Baths of Apoquinto—Slopes of the Cordillera—Excursion to Santa Rosa de los Andes and the valley of Aconcagua—Return to Valparaiso—Voyage in the German steamer *Rhamses*—Visit to Lota—Parque of Lota—Coast of Southern Chili—Gulf of Peñas—Hale Cove—Messier's Channel—Beautiful scenery—The English narrows—Eden harbour—Winter vegetation—Eyre Sound—Floating ice—Sarmiento Channel—Puerto Bueno—Smyth's Channel—Entrance to the Straits of Magellan—Glorious morning—Borya Bay—Mount Sarmiento 188

CHAPTER V.

Arrival at Sandy Point—Difficulties as to lodging—Story of the mutiny—Patagonian ladies—Agreeable society in the Straits of Magellan—Winter aspect of the flora—Patagonians and Fuegians—Habits of the South American ostrich—Waiting for the steamer—Departure—Climate of the Straits and of the southern hemisphere—Voyage to Monte Video—Saturnalia of children—City of Monte Video—Signor Bartolomeo Bossi; his explorations—Neighbourhood of the city—Uruguayan politics—River steamer—Excursion to Paisandu—Voyage on the Uruguay—Use of the telephone—Excursion to the camp—Aspect of the flora—Arrival at Buenos Ayres—Industrial Exhibition—Argentine forests—The cathedral of Buenos Ayres—Excursion to La Boca—Argentaria as a field for emigration ... 248

CHAPTER VI.

Voyage from Buenos Ayres to Santos—Tropical vegetation in Brazil—Visit to San Paulo—Journey from San Paulo to Rio Janeiro—Valley of the Parahyba do Sul—Ancient mountains of Brazil—Rio Janeiro—Visit to Petropolis—Falls of Itamariti—Struggle for existence in a tropical forest—The hermit of Petropolis—Morning view over the Bay of Rio—A gorgeous

CONTENTS.

PAGE

flowering shrub—Visit to Tijuca—Yellow fever in Brazil—A giant of the forest—Voyage to Bahia and Pernambuco—Equatorial rains—Fernando Noronha—St. Vincent in the Cape Verde Islands—Trade winds of the North Atlantic—Lisbon—Return to England 303

APPENDIX A.—On the fall of temperature in ascending to heights above the sea-level 369

APPENDIX B.—Remarks on Mr. Croll's theory of secular changes of the earth's climate 393

NOTES OF A NATURALIST
IN SOUTH AMERICA.

CHAPTER I.

Voyage across the Atlantic—Barbadoes—Jamaica—Isthmus of Panama—Buenaventura, tropical forest—Guayaquil and the river Guayas—Payta—The rainless zone of Peru—Voyage to Callao.

A VOYAGE across the Atlantic in a large ocean steamer is now as familiar and as little troublesome as the journey from London to Paris. It rarely offers any incident worth recounting, and yet, especially as a first experience, it supplies an abundant variety of sources of curiosity and interest. It is easy for a man to sit down at home and within the walls of his own study to find the requisite materials for investigating the still unsolved problems presented by the physics and meteorology of the ocean, or the evidence favourable or hostile to the important modern doctrine of the permanence of the great ocean valleys; but in point of fact very few men who stay at home do occupy themselves with these questions, and it is no slight privilege to feel drawn towards them by the hourly suggestions received during a sea-voyage. Nor is it possible to make light of the simpler pleasures caused by the satisfaction of mere curiosity, when that is linked by association with the pictures on which the fancy has worked from one's earliest childhood onward. The starting of a covey of flying-fish, the fringe of cocos palms rising against the horizon, the Southern Cross and the Magellanic clouds, the reversed apparent motion of the sun from right to left—none of them very marvellous as mere observed facts—are so many keys that unlock the closed-up recesses, the blue chambers of the memory, which the youthful imagination had peopled with shapes of beauty and wonder and mystery.

Some thrill of delightful anticipation was, I presume, felt by many of the passengers who went on board the royal mail steamer *Don* in Southampton Water on the 17th of March, 1882. Amid the usual waving of handkerchiefs from the friends who remained behind on board the tender, we glided seaward, and by four p.m. were going at half speed abreast of the Isle of Wight. The good ship had suffered severely during the preceding winter on her homeward passage from the West Indies, when the heavy seas which swept her upper deck had carried away the covering of her engine-room, stove in the chief officer's cabin, and severely injured her commander, Captain Woolward. On this occasion our voyage was easy and prosperous, and nothing occurred to test severely the careful seamanship of Captain Gillies, who had taken the temporary command.

On the 19th the barometer, which, in spite of a gentle breeze from south-west, had stood as high as 30·40, fell about a quarter of an inch between sunrise and sunset; and in the night, on the only occasion during the entire voyage, remained for some hours below 30·00. A moderate breeze from the north brought with it a disproportionately heavy sea, and although there was no sensible pitching, the ship rolled so heavily as to send many of the passengers to solitary confinement in their berths. This continued throughout the 20th, afterwards styled Black Monday by the sufferers from sea-sickness, and we escaped into smoother water only on the evening of the following day. The discomfort which I felt from fancying that I had "lost my sea legs" was entirely relieved by fortunately coming across a distinguished naval officer, on his way to take a command on the West Indian station, who like myself was forced to hold on with both hands during the rolling of the ship.

It was clear that we had passed at no great distance from a cyclone in the North Atlantic—one of those disturbances whose visits are so often predicted from the western continent, but which so often fortunately lose their way or get dissipated before they approach our shores. It would seem that little progress has been made in forecasting the direction in which these great aërial eddies traverse the ocean, or the conditions under which they expend their force. It seems allowable to suppose that the most important of the causes influencing their direction depend upon the general movements of the great currents of the atmosphere; and that, as these are constantly modified by the changing position of the earth in her orbit, the element of season is primarily to be considered. It being admitted that the origin of these disturbances is to be sought in the abnormal heating or cooling of some considerable portion of the earth's surface, it would seem that, in the case of the Atlantic, local causes can have little effect, unless we suppose that the heating of the surface of the Azores in summer, or the annual descent of icebergs from the polar seas, are adequate to influence the march of a travelling cyclone.

On the evening of the 20th the barometer had risen again to its former position, rather over 39.40 inches; the mean of the four following days was 30.55, and that of the entire run from Southampton to Barbadoes was 30.36. This fact of the continuance of high or low pressures at the sea-level at certain seasons in some parts of the world has scarcely been sufficiently noted in connection with the ordinary rules for the measurement of heights by means of the barometer. The tables supplied to travellers are all calculated on the assumption that the pressure at the sea-level is constant—the English tables fixing the amount at 30.00 inches of mercury, those calculated on the continent starting from a pressure of 760 millimetres, or about 29.921 inches. It is admitted that this mode of determining heights, when comparative observations at a known station are not available, is subject to serious unavoidable error. With regard, however, to mountains not remote from the sea-coast, it may be possible to lessen this inconvenience in many parts of the world by substituting for the assumed uniform pressure that higher or lower amount which is known to prevail at given seasons. Such a correction could not, of course, be made available in very variable climates, such as that of the British Islands, but might be applied in many parts of the broad zone lying within 40° of the equator.

Soon after ten p.m. on the 21st we were abreast of the bright light which marks the harbour of St. Michael's, but, the night being dark, we saw very little of that or any other of the Azores group. The spring temperature of these islands is about the same as that of places in the same latitude in Portugal; but it appears that the cooling effect of the east and north-east winds prevailing at that season must in the mid-Atlantic extend even much farther south. With generally fair settled weather, the thermometer rose very slowly as we advanced towards the tropics. Between the 18th and 24th of March, in passing from 50° to 29° north latitude, the mean daily temperature rose only from about 55° to about 65° Fahr.—the thermometer never rising to 70°, nor falling below 52°. Notwithstanding the relatively low temperature, a few flying-fish were seen on the 24th—rare, it is said, outside the tropics so early in the year, though sometimes seen in summer as far north as the Azores.

On March 25th we, for the first time, became conscious of a decided though moderate change of climate. The thermometer at noon stood at 71°, and was not seen to fall below 70° until, some three weeks later, off the Peruvian coast, we met the cold antarctic current which plays so great a part in the meteorology of that region. We were now in the regular track of the north-east trade wind, and my mind was somewhat exercised to account for the circumstance, said to be of usual occurrence, that the breeze increases in strength from sunrise during the day, and falls off, though it does not die away, towards nightfall. It is easy to understand the cause of this intermittence in breezes on shore, whether near the sea-coast or in the neighbourhood of mountain ranges, inasmuch as their direction and strength are determined by the unequal heating of the surface; but the trade winds form a main part of the general system of aërial circulation over the surface of our planet, and, supposing the phenomenon to be of a normal character, the explanation is not quite simple. Regarding the trade wind as a great current set up in the atmosphere, it is conceivable that the heating and consequent expansion which must occur as the sun acts upon it, tends to increase the rate of flow at the bottom of the aërial stream, while the cooling which ensues as the sun's heat is withdrawn, has the contrary effect.

On this and the next day or two my attention was called to the frequent recurrence of masses of yellow seaweed, sometimes in irregular patches, but more frequently arranged in regular bands, two or three yards in width, and extending in a straight line as far as the eye could reach. We were here at no great distance from the great sargassum fields of the Northern Atlantic, but I was unable to satisfy myself that the species seen from the steamer was that which mainly forms the sargassum beds; and, whatever it might be, this arrangement in long straight strips seemed deserving of further inquiry. More flying-fish were now seen, and two or three small whales of the species called by seamen "black-fish" were sighted during this part of the voyage.

On the afternoon of the 26th we entered the tropics, and this and the following day were thoroughly enjoyable, but did not offer much of novelty. The colour of the sea was here of a much deeper and purer blue (rivalling that of the Mediterranean) than we had hitherto found it, while that of the sky was much paler. The light *cumuli* with ill-defined edges were such as we are used to in British summer weather; and, excepting that the interval of twilight was sensibly shorter, the sunsets were devoid of special interest. At this season the Southern Cross was above the horizon about nightfall, and was made out by the practised eyes of some of the officers; but, in truth, it remains a somewhat insignificant object when seen from the northern side of the equator, and to enjoy the full splendour of that stellar hemisphere one must reach high southern latitudes.

Although the thermometer never quite reached 80° Fahr. in the shade until we touched land, the weather on the 28th and 29th was hot and close, and few passengers kept up the wholesome practice of a constitutional walk on the long deck of the *Don*. Of the rain which constantly seemed impending very little fell.

At daybreak on the morning of the 30th, in twelve days and seventeen hours, we completed the run of about 3340 nautical miles which separates Southampton from Barbadoes, and found ourselves in the roads of Bridgetown, about a mile from the shore.

Being somewhat prepared, I was not altogether surprised to find that this first view of a tropical island forcibly reminded me of the last land I had beheld at home—the northern shores of the Isle of Wight. Long swelling hills, on which well-grown trees intervene between tracts of tillage, present much the same general outline, and at this distance the only marked difference was the intense dark-green colour of the large trees that embower the town and nearly conceal all but a few of the chief buildings. The appearance of things as the morning advanced quite confirmed the reputation of this small island as the most prosperous, and, in proportion to its extent, the most productive of the West Indian Islands. With an area not greater than that of the Isle of Wight, and a population of about sixty thousand whites and rather more than a hundred thousand negroes, the value of the exports and imports surpasses a million sterling under each head; and, besides this, it is the centre of a considerable transit trade with the other islands. Under local representative institutions, which have subsisted since the island was first occupied by the English early in the seventeenth century, the finances are flourishing, and the colonial government is free from debt. The average annual produce of sugar is reckoned at forty-four thousand hogsheads, but varies with the amount of rainfall. This averages from fifty-eight to fifty-nine inches annually, but any considerable deficiency, such as occurred in the year 1873, leads to a proportionate diminution in the sugar crop.

Among other tokens of civilization, the harbour

police at Bridgetown appeared to be thoroughly efficient. As, about nine o'clock, we prepared to go ashore, we found on deck two privates—black men in plain uniform—who seemed to have no difficulty in keeping perfect order amid the crowd of boatmen that swarmed round the big ship. We had already learned the event of the hour—the fall of three inches of rain during the day and night preceding our arrival. This is more than usually falls during the entire month of March, and seemed to be welcomed by the entire population. On landing we encountered a good deal of greasy grey mud in the streets, but all was nearly dry when, after a short excursion, we returned in the afternoon. After a short stay in the town, where there was a little shopping to be done, and where some of my companions indulged in a second breakfast of fried flying-fish, I started with a pleasant party of fellow-travellers to see something of the island. It was arranged that, after a drive of six or seven miles, we should go to luncheon at the house of Mr. C——, the owner of a sugar-plantation, whose brother, Colonel C——, was one of our fellow-passengers. We enjoyed the benefit of the recent heavy rain in the comparative coolness of the air—the thermometer scarcely rose above 80° Fahr. in the shade—and in freedom from dust.

A small, low island, nearly every acre of which has been reduced to cultivation, cannot offer very much of picturesque beauty; nevertheless the first peep of the tropics did not fail to present abundant matter of interest. In this part of the world the dry season, now coming to an end, is the winter of vegetation, and, of course, there was not very much to be seen of the herbaceous flora ; but the beauty of the trees and the rich hues of their foliage quite surpassed my anticipations. The majority of these are plants introduced either from the larger islands or from more distant tropical countries, that have been planted in the neighbourhood of houses.

One of the first that strikes a new-comer in the tropics is the mango tree, which, though introduced by man from its original home in tropical Asia, is now common throughout the hotter parts of America. Its widespreading branches, bearing dense tufts of large leathery leaves, make it as welcome for the sake of protection from the sun as for its fruit, which is a luxury that some persons never learn to appreciate. The cinnamon tree (*Canella alba*), common in most of the West Indian Islands, is another of the plants that serve for ornament and shade while ministering products useful to man. Of the smaller shade-trees, the pimento (probably *Pimenta acris*) was also conspicuous, and very many others which I failed to recognize, might be added to the new impressions of the first day in the tropics. One of the most curious is that known to the English residents as the sand-box tree, the *Hura crepitans* of botanists. It belongs to the *Euphorbiaceæ*, or Spurge family, but is strangely unlike any of the Old-World forms of that order. Here the fruit is in form rather like a small melon, of hard woody texture, divided into numerous—ten to twenty—cells. If, when taken from the tree, the top is sawn off and the seeds scooped out, no farther change occurs, and it may be, and often is, as the name implies, used as a sand-box. But if left until the seeds are mature, the whole capsule bursts open with a loud report, scattering the seeds to a distance. Thinking that a small young fruit, if dried very gradually, might escape this result, I carried one away, which, after my return to Europe, I placed in a small wooden box in my herbarium. Some nine months after it had been collected it must have exploded in my absence, for, unlocking the room one day, I found the box broken to pieces, and the valves of the fruit and the seeds scattered in all directions about the room.

Next to the vegetable inhabitants, I was interested in the black population of the island. The first impression on finding one's self amid fellow-creatures so markedly different in physical characters is one of strangeness, and one is tempted to ask whether, after all, there can be any pith in the arguments once confidently urged to establish a specific difference between the negro and the white man. But this very quickly wears away, and a contrary impression arises. The second thought is that, considering what we know of the conditions under which the native races of Equatorial Africa have been developed during an unrecorded series of ages, and of the subsequent conditions during several generations of slavery, the surprising thing is that the differences should not be far greater than they are.

It would be very rash to draw positive conclusions from what could be seen in a visit of a few hours, but, undoubtedly, the general effect was pleasing, and tended to confirm the assertion that the difficult problem of converting a population of black slaves into useful members of a free community has been better solved in Barbadoes than in any other European colony. So far as the elementary wants are concerned, there was a complete absence of the painful suspicion so commonly felt as regards the poor in Europe and the East, that their food is either insufficient or unwholesome. With very few exceptions they all seemed sleek and well fed, and their clothing showed no symptoms of poverty. In the town their dress was generally neat, and most of the women made a display of bright colour in handkerchiefs and parasols. What struck me most was a general air of good humour and enjoyment. One may be misled in this respect by the facial characteristics of the black race, which, in the absence of disturbing causes, readily turn to a smile or a grin. But, whether in the streets of Bridgetown or botanizing among the fields in the country, and using the few opportunities of speaking to the people, the same impression was retained.

Their manner in speaking to whites seemed to imply neither servility nor yet the independence which characterizes the Arab or the Moor. A latent sense of inferiority seemed to be combined with a complete absence of shyness or apprehension, as in children used to kind treatment, and not too carefully drilled. We happened to halt near a spot where there was a cluster of labourers' cabins, and a school well filled with small children. There had been a wedding in Bridgetown that morning, and as we halted two carriages passed, carrying the bridal party to some house in the country. All the inhabitants rushed out at once, and contended, young and old, in the most boisterous cheering. Perhaps this meant little more than the mere love of noise, as when boys cheer a passing railway train, but it argued, at least, the absence of any feeling of race animosity.

The houses of the labouring population, whether in town or country, are mere sheds, seemingly of the frailest materials, the walls of thin upright boards, and roofed with small imbricated wooden shingles, such as one sometimes sees in Tyrol ; but there must be a very substantial framework, or they would be annually carried away by the August hurricanes. The interiors appeared to be fairly clean, and in a country where cold is unknown good houses are luxuries, not necessaries of life.

One need not go far to seek the explanation of the superior condition of Barbadoes as compared with the other West Indian Islands. Unlike these, there was here no waste land ; every acre was occupied, and the emancipated negro could not follow the very natural but unfortunate instinct which elsewhere led him to squat in idleness, supporting life on a few bananas and other produce that cost but a few days' labour in the year. Apart from this, it is said that the Barbadoes, unlike the Jamaica, planters showed practical intelligence in at once recognizing the new conditions created by the Act of Emancipation, and, by offering fair wages and giving their personal influence and supervision, helping to convert the slave into an industrious freeman. Whatever poets may have fancied of the delights of lotus-eating, it seems to be true in the tropics, as well as in temperate climates, that there is more contentment and real enjoyment of life among people who are held to regular daily work—not excessive or exhausting—than among those who have little or nothing to do.

The house at which we were hospitably entertained, with no architectural pretensions, struck us as admirably suited to the climate. On the ground floor, several spacious and airy sitting-rooms opened on a broad verandah that ran round the building, and a number of fine trees close at hand, with the dense impervious foliage characteristic of the tropics, offered the alternative of sitting in the open air. One of the natural advantages of Barbadoes is the almost complete absence of noxious and venomous insects and reptiles. The frequency of poisonous snakes in some of the islands, especially Martinique and Sta. Lucia, must seriously interfere with the pleasures of a country life.

The voyage from Barbadoes to Jacmel, which occupied the greater part of three nights and two days, was highly enjoyable, but uneventful. With a temperature of about 80° in the shade, and a pleasant breeze from the north-east, life on deck was much more attractive than any occupation in the cabins, and nothing more laborious than reading an interesting book, such as Tschudi's "Travels in Peru," or at the utmost some brushing up of nearly forgotten Spanish, could be undertaken. In the early morning, the rising of the coveys of flying-fish as the steamer disturbed them from their rest on the surface, with their great silvery fins glancing in the level rays of the sun, was always an attractive sight. They certainly often change the direction of their flight as they momentarily touch the surface, but I could not satisfy myself whether this depended on a muscular effort of the animal, or merely on the angle at which it happened to strike the irregular surface of the little dancing waves that surrounded us.

About sunrise on the 2nd of April the anchor was let go, and we found ourselves in the harbour of Jacmel, the only port on the south side of the great island of Hayti. The Royal Mail steamers call here periodically to deliver letters and to receive a bag which, after due fumigation and such other incantations as are deemed proper, is delivered at the end of a long pole. The entire island being supposed to be constantly subject to zymotic diseases, especially small-pox which is the great scourge of the negro race, no further communication with the shore is permitted, and within less than two hours we were again under way. The hills surrounding the harbour are apparently covered with forest, the trees being of no great size, but of the most brilliant green; but I could detect no dwellings of a superior class such as Europeans would be sure to construct in picturesque and healthy spots near a seaport. As we ran for more than twenty miles very near the coast, I could at first detect here and there small patches of cleared ground with sheds or huts; but beyond the distance of a few miles these ceased, and no token of the presence of man was discernible.

Making large allowance for exaggeration, and having had the opportunity of correcting some loose reports by the more careful and accurate information afterwards received from a gentleman who resided for some time at Port au Prince as the representative of a European power, it is impossible for me to avoid the conclusion that, in the hands of its black possessors, this noble island has retrograded to a condition of savagery little, if at all, superior to that of the regions of tropical Africa whence they originally came.

There may be but slight foundation for the reports as to the revival of cannibal customs in the interior of the island; but it would seem that the sanguinary encounters so frequently recurring between the people of the rival republics between whom the island is divided, differ little in point of ferocity from those of Ashantee or Dahomey. The political institutions, caricatures of those of the United States, have produced in astonishing luxuriance all the abuses characteristic of different types of misgovernment, and the few men distinguished by superior intelligence and a desire for rational progress have sought in vain for support in efforts for reform. The condition of the two republics, Hayti and San Domingo, seems to be the *reductio ad absurdum* of the theories which ascribe to free institutions an inherent power of promoting human progress.

April 3 was a day to be long remembered. Barbadoes to Jamaica is as Champagne or Mecklenburg compared to Switzerland or Tyrol, and now for the first time the dream of tropical nature became a reality. At six p.m. we passed Port Royal, and about seven had cast anchor at Kingston. The first impression on landing here is unfavourable. The buildings are mean, the thoroughfares and side-paths out of repair, the people in the streets seem to have nothing to do and to be doing it, the general air that of listlessness and neglect. Altogether the place contrasts disadvantageously with the ports of Spanish America, to say nothing of our own colonies. But Kingston was not to detain us, and the overpowering attraction was towards the range of the Blue Mountains, on which my eyes had been fixed all the morning as we approached the shore. We were told that we must return to the ship at five o'clock, so that it was hopeless to attempt to reach even the middle zone of the mountains, and all that could be done with advantage was to engage a carriage to a place called Gordontown, in a valley which is the ordinary route to Newcastle and other places in the mountains. After a delay which to our impatience seemed unreasonable, I started in a tolerable carriage with W——, an old friend who was proceeding to Lima as commissioner from the Court of Chancery to receive evidence in an important pending lawsuit, and who, although not a naturalist, gave effective and valuable help on this and other subsequent occasions in the work of plant-collecting.

For a distance of four or five miles the land slopes very gently from the coast towards the roots of the hills. This tract is partly occupied by sugar-plantations; but our road lay for some time among small country houses, each surrounded by pleasure-ground or garden. As the dry season was not yet over, the country here looked parched; but I saw many trees and shrubs new to me, many of them laden with flowers, and found it hard to keep my resolution not

C

to stop the carriage until we should reach Gordontown. The excitement increased as we entered the valley, and the road began to wind up the slopes above the right bank of the torrent, where at every yard some new object came into view. It was near eleven a.m. when we reached the little inn, which, with four or five houses, make the station of Gordontown, where the carriage road ends, and horses are hired by those bound for Newcastle or other places in the hills. No time was to be lost, and we were speedily on our way to ramble up the valley, keeping as near as might be to the banks of the torrent.

The first effect upon one accustomed only to the vegetation of the temperate zone is simply bewildering. As I expressed it at the time, it seemed as if the inmates of the plant-houses at Kew had broken loose and run scrambling up the rocky hills that enclose the valley. These are of a red arenaceous rock, rough and broken, but affording ample hold for trees as well as smaller plants. The torrent at this season was shrunk to slender dimensions, but is never wholly dry; and I was somewhat surprised to find that on the steep slopes exposed to the full sunshine the vegetation was much less parched than one commonly finds it in summer in the Mediterranean region, and even to gather a good many ferns on exposed banks. It would appear that, even in the dry season, the air must here be nearly saturated with aqueous vapour, and that abundant dews must supply the needs of delicate plants. Not many species were in flower, but yet there was more than sufficient to occupy the short time available. *Malvaceæ* and *Con-*

volvulaceæ were the most prominent forms; but to a new-comer the most lively interest attaches to groups never before seen in a wild state, such as *Passiflora*—of which two species were found in flower—a first solitary representative of the great tropical American family of *Melastomaceæ*, or the gorgeous Amaryllid, *Hippeastrum equestre*, hiding in shady places by the stream.

Although Gordontown can scarcely be so much as a thousand feet above the sea-level, the climate is very sensibly cooler than that of Kingston. When we left the town the thermometer stood at 83° in the shade, while here at midday the sea-breeze felt positively cold, and I was glad to have with me an extra garment. A light luncheon of ham and eggs, with guava sweetmeat for dessert, was soon despatched; and, as I wished to halt at several spots on the way, we started about half-past two, laden with the spoils of the excursion, and reached the steamer before five o'clock. Great was my disgust to find that there was no intention of starting until nine a.m. the next morning, and this was changed to indignation when it came to be known that we had been deprived of the priceless pleasure of a trip to the mountains by the deliberate misstatement of the company's superintendent, who had arranged to embark on the following morning three hundred negroes going to work on the Panama Ship Canal.

A stranger can scarcely fail to observe a marked difference between the negro population of Jamaica and that of Barbadoes. In the larger island, while no way deficient in physical qualities, they appear decidedly inferior in intelligence, activity, and courtesy towards their white neighbours. It is said that the independent class, who live by cultivating small patches of land on which they have squatted, has of late years much improved, and that the increasing desire for purchasable comforts and luxuries has begun to develop habits of steady industry; but as regards the mass of the people who live by wages, there are many indications of a sullen dislike towards the descendants of their former masters which some trifling provocation may at any time inflame to a pitch of wild ferocity. Some who have lived in the island maintain that a general rising with a view to the massacre of the white population is not an impossible occurrence, and, however improbable it may appear, there is ample reason for constant vigilance on the part of those responsible for the government of the island. Such vigilance, it must be remembered, is quite as much requisite to prevent acts of real or apparent injustice towards the inferior race, as to repress the first beginnings of violence if some spark should fire the mine of suppressed hatred.

After a too short visit to this beautiful island, we were under way before ten a.m. on April 4th, and before midday the outline of the Blue Mountains of Jamaica was fast fading in the northern horizon. Throughout the greater part of the run from Kingston we encountered a moderately brisk breeze, which gradually veered from south-east to south-west, and this, according to our experienced captain, commonly occurs at this season. It may be conjectured that the great mountain barrier extending on the south side of the Caribbean Sea through Venezuela and Colombia deflects the current of the north-east trade wind until it finally flows in an exactly contrary direction. Whatever its origin may be, it might be supposed that the interference of a current from the south-west with the course of the regular trade wind would give rise to storms of dangerous violence. These, however, rarely if ever occur during the spring months. It may be that, on the meeting of contrary currents of unequal temperature, the ordinary result is that the warmer current rises and flows over the cooler one without actual interference.

Before sunrise on the morning of the 6th we reached Colon, and, after a little inevitable delay, took leave of our excellent commander, and set foot on the American continent at a spot which seems destined to become familiar to the civilized world as the eastern termination of the Panama Ship Canal. People who love to paint in dark colours had done their best to make us uncomfortable as to the part of the journey between the arrival at Colon and the departure from Panama. The regular train crossing the isthmus starts very early from Colon, and we should be forced to remain during the greater part of the day breathing the deadly exhalations of that ill-famed port. In point of unhealthiness Panama is but little better than Colon, and as the weekly steamer of the Pacific Navigation Company bound southward would have departed one or two days before our arrival, we were sure to be detained for five or six days, equally trying to the health and temper. Fully believing these vaticinations to be much exaggerated, we had no opportunity of testing them. A free use of the telegraph on the morning of our arrival at Jamaica, and the courtesy of the officials of the various companies concerned, relieved us from all anxiety, and reduced our stay within the shortest possible limits. It was true that the regular train had been despatched before we could land, but a special engine was in readiness to convey us across the isthmus, and the agent for the Pacific mail steamer at Panama had detained the ship bound for Lima until the same evening in order to enable us to continue our voyage.

Since the commencement of the works connected with the canal, Colon must have undergone much improvement. The bronze statue of Columbus presented by the Empress Eugénie, which for many years had lain prostrate in the mud of the sea-beach, has been cleansed and placed upon a stone pedestal. A number of stores, frail structures of wooden planks, were arranged in an irregular street, and displayed a great variety of European goods. It was rather surprising to find the prices of sundry small articles purchased here extremely moderate. One might suppose that the only inducement that could lead people to trade in a spot of such evil repute would be the hope of exorbitant profits enabling them soon to retire from business.

Of the works connected with the Ship Canal little was to be seen from the railway cars. For its eastern termination the mouth of the Chagres river, which reaches the sea close to Colon, has been selected. I am not aware whether it is proposed to divert the course of that stream from the channel of the canal, but, to judge from the appearance of its banks and the extensive mangrove swamps on either side, it appears to bear down a great amount of fine alluvial mud, which, if discharged into the canal, must be a source of future difficulty. What chiefly struck the eye of the passing traveller was the broad band which had been cleared across the isthmus to mark the line of the future canal. It is fully a hundred metres in width, and seemingly carried in a nearly straight line through the forest and over the hills that lie on the western side near to Panama. This clearing does not appear a very serious undertaking, but in a region where the energy of vegetation is so marvellous, must have cost an immense amount of labour, and to keep the line open, if that be found expedient, will demand no small yearly expenditure. There is here, properly speaking, no dry season. The rains recur at frequent intervals throughout the year, and to keep back the ever-encroaching sea of vegetation the axe is in constant requisition.

In the interest of the human race, it is impossible not to desire the success of the Ship Canal, but it must not be forgotten that the project is of a character so gigantic that all previous experience, such as that of the Suez Canal, fails to give a measure of the difficulties to be encountered, or of the outlay required to overcome them. Engineers may doubtless calculate with sufficient accuracy the number of millions of cubic yards of rock or earth that must be removed, and may estimate approximately the cost of labour and materials; but the obstacles due to the climate and physical conditions of this region are a formidable addition whose amount experience alone can fully determine. The only race combining physical strength with any moderate adaptation to the climate is apparently the African negro, and even with these the amount of sickness and mortality is said to be alarmingly great. The field from which negro labour can be recruited, though large, is by no means unlimited, and it is to be expected that the rate of wages must be considerably increased as time advances. The conditions of the problem have no doubt been carefully studied by the remarkable man to whom its existence is due, and by the able assistants whom he has consulted; but it may not be too rash to hazard the prediction that, apart from any international difficulties, its success may depend upon the more or less complete realization of two desiderata—first, the extensive application of labour-saving machinery, for which perhaps the heavy rainfall may supply the motive power; secondly, the possibility, by completely clearing the summits of some of the higher hills near the line, of establishing healthy sites whence workmen could be conveyed to the required points during the day and brought back before nightfall.

Nothing in our brief experience suggested the idea of an especially unhealthy region, and the feelings of a botanist at being whirled so rapidly through a land teeming with objects of curiosity and interest are better imagined than expressed. For more than half the distance the line is simply a trench cut through the forest, which is restrained from invading and burying the rails only by constant clearing on either side. The trees were not very large, but seemed to include a vast variety of forms. More striking were the masses of climbers, parasites, and epiphytes, to say nothing of the rich and strange herbaceous plants that fringed the edge of the forest. Our train, being express, gave but a single chance of distinguishing anything amid the crowd of passing objects—during a brief halt at a station about half-way across the isthmus, round which was a cluster of small houses or huts, inhabited by Indians. Their features were much less remote from the European type than I had expected—less remote, I thought, than those of many Asiatics of Mongol stock. Ten minutes on the verge of the surging mass of vegetation that surrounded us gave a tantalizing first peep at the flora of Equatorial America. Many forms hitherto seen only in herbaria or hot-houses—several *Melastomaceæ, Heliconia, Costus*, and the like—were hastily gathered ; but the summons to return to the train speedily calmed the momentarily increasing excitement. Although the sky was almost completely free from clouds, and the sun very near the zenith, the heat was no way excessive. My thermometers had been stowed away in the hurry of leaving the steamer, but I do not believe that the shade temperature was higher than 84° Fahr. On the western side of the isthmus the land rises into hills some five or six hundred feet in height, and between these the railway winds to the summit level, thence descending rather rapidly towards Panama. What a crowd of associations are evoked by the first view of the Pacific! What trains of mental pictures have gathered round the records of the early voyagers, the adventurers, the scientific explorers! Strangely enough, the most vividly impressed on my memory was a rough illustration in a child's book, given to me on my seventh birthday, representing Vasco Nuñez, as, from the summit of the ridge of Darien, he, first of all western men, cast his wondering eyes over the boundless, till then unsuspected, ocean. He has climbed the steep shattered rocks, and, as he gains the crest of the ridge, has grasped a projecting fragment to steady himself on the edge of the dizzy declivity. Even now, after looking on the gently swelling hills, so completely forest-covered that without extensive clearing a distant view would be impossible, I find it hard to believe that that picture does not represent some portion of my actual past experience.

I do not know whether, in connection with the vivid recollection either of actual scenes or illustrations dating from early life, attention has been sufficiently called to the curious tricks which the brain not seldom performs in discharging its function of keeper of the records. In my experience it is common to find, on revisiting after many years a spot of which one believes one's self to have a vivid and accurate recollection, that the mental picture has undergone some curious changes. The materials of the scene are, so to say, all present, but their arrangement has been unaccountably altered. The torrent, the bridge, the house, the tree, the peak in the background, are all there, but they are not in their right places. The house has somehow got to the wrong side of the torrent, or the peak rises on the right of the tree instead of the left. A picture vividly retained in the mind is one that has been frequently recalled to memory. If at any time, when it has been long dormant, the actual recollection has become somewhat imperfect, the imagination fills up by an effort the incomplete portion. When next summoned by some train of association, the image present to the mind is no longer the original picture, but the altered version of it in the state in which it was left after being last retouched.

In about four hours from Colon we reached the Panama terminus, and found a large waggonette, or roofless omnibus, waiting to convey us to the Grand Hotel. A pair of small ragged horses, rushing at a canter down the steep slopes and scrambling up on the other side over the rough blocks that form the pavement, made our vehicle roll and jolt in a fashion that would have disquieted nervous passengers. It would be difficult to find elsewhere in the world a stranger assemblage than that to be found at the Grand Hotel of Panama. The ground floor, with several large rooms, is occupied day and night for eating, drinking, smoking, and loud discussion by the floating foreign population of the town. At the present time the engineers and other officials connected with the Ship Canal formed the predominant element; but, along with a sprinkling of many other nationalities, the most characteristic groups consisted of refugees from all the republics of Central and South America, who find substantial reasons for quitting their homes, and who resort to Panama as a sanctuary whence some new turn in the wheel of revolution may recall them to some position of distinction and profit.

We were fortunate in having in our company Mr. W———, a gentleman of Polish descent, to whose lively conversation we had owed much information and amusement during the voyage from Southampton. Now the owner of a large estate in Ecuador, he had long known this region, and appeared to be on terms of familiar acquaintance with all the strange visitors gathered in the saloons at Panama, from the ex-President of Peru to the negro head-waiter. The latter, as we learned, was not the least important member of the assemblage. In one of the numerous revolutions at Panama he had played a leading part, and had attained the rank of colonel. His party being then out of office, he had for the time returned to private life, but may possibly at the present day be again an important person in the state.

For the first time since leaving England the heat at Panama during the midday hours was felt to be oppressive, and we were content with a short stroll, which, to any one familiar with old Spain, offered little novelty. Unlike such mushroom spots as Colon, Panama has all the appearance of an old Spanish provincial town. It has suffered less from earthquakes than most of the places on the west coast, and a large proportion of the buildings, including a rather large cathedral, remain as they were built two or three centuries ago.

As the anchorage for large steamers is about three miles from the town, we had an early summons to go on board a small tender that lay alongside of a half-ruined wharf, but were then detained more than an hour, for no apparent reason other than as a tribute to the habits of the population of this region. The time was not wholly wasted, as even the least observant passengers were struck with admiration at the performances of a swarm of small birds, many hundreds in number, that seemed to have selected the space over the shallow water opposite the town for their evolutions. For more than half an hour they continued to whirl in long loops or nearly circular sweeps, with no other apparent motive than the pleasure of the exercise. Seen from a distance, the appearance was that of a wreath; nearer at hand, the arrangement was seen to be constantly varying. Sometimes the birds were so close together that it seemed as if their wings must jostle; sometimes they were drawn out into long curves, looking silvery white when the sun fell upon their breasts, and of a darker tint at other incidences. Mr. W—— asserted that the bird is a kind of snipe, but I have no doubt that it is a tern.

At last the little tender glided from the wharf, and for the first time we gained a general view of the town, which has a full share of that element of picturesqueness which is so strangely associated with decay. The old ramparts fast crumbling away, here and there rent by earthquakes, and backed by time-stained buildings, would offer many a study to the painter. Sunset was at hand when we reached the steamer *Islay*, anchored under the lee of one of the small islands of the bay, and were fortunate in finding among the not too numerous passengers several whose society added to the interest of the voyage.

One of the effects of the habitual use of maps on a small scale is that untravelled persons, even though conversant with the facts of geography, feel it difficult to realize the great dimensions of the more distant parts of the world as compared with our diminutive European continent. Thus it came on me with something of surprise that the Bay of Panama is fully a hundred and twenty sea miles across from headland to headland, and that the run from Panama to Callao, which is scarcely one-third of the length of the South American continent, is rather longer than that from Bergen to the Straits of Gibraltar. The case, of course, is much worse with those accustomed to use maps on Mercator's projection. It profits nothing to explain, even to the most intelligent youth, the nature and amount of the errors involved in that mode of representing a spherical surface on a plane. I verily believe that all the mischief done by the stupidity, ignorance, and perversity of the writers of bad schoolbooks is trifling compared to the amount of false ideas spread through the world by the productions of that respectable Fleming.

The steamers of the Pacific Mail Company employed for the traffic between San Francisco and Valparaiso are as perfectly suited to the peculiar conditions of the navigation as they would be unfit for long sea-voyages in any other part of the world. In the calm waters of this region, rarely ruffled even by a stiff breeze, the fortunate seamen engaged in this service know no hardships from storm or cold. Their only anxiety is from the fogs that at some seasons beset parts of the coast. In each voyage they pass under a vertical sun, but the air and the water are cooler than in any other part of the equatorial zone; and all that is needed for their physical comfort, and that of their passengers, is free ventilation and shade from the sun. These desiderata are fully secured. The main-deck is open to the air, and the steerage passengers, who are encamped amidships and on the fore-deck, are satisfied at night with the amount of privacy secured by hanging some piece of stuff to represent a curtain round each family group. On the upper deck are ranged the state rooms of the first-class passengers, each with a door and window opening seaward. Above this, again, a spar-deck carried flush from stem to stern affords ample opportunity for exercise, and is itself sheltered from the sun by an awning during the hot hours. In such conditions, where merely to breathe is to enjoy, the only danger is that of subsiding into mere lotus-eating. From this I was fortunately preserved by the rather troublesome task of drying in satisfactory condition the plants which I had hastily gathered in Jamaica and in crossing the isthmus.

I had supposed that the distinctly green colour of the water in Panama Bay, so different from the blue tint of the open Atlantic, might be due to some local peculiarity; but on the following day, April 7, while about a hundred miles from land, I observed that the same colour was preserved, and I subsequently extended the observation along the coast to about 5° south, where we encountered the antarctic current. Farther south I should describe the hue of the water as a somewhat turbid dark blue, reminding one of the water of the North Atlantic as seen in approaching the British Islands.

At daybreak on April 8 we found ourselves approaching the port of Buenaventura. Long before it was possible to land I was ready, thrilling with interest and curiosity respecting a region so entirely new—an interest enhanced, perhaps, by the extent of ignorance of which I was inwardly conscious. Knowing this place to be the only port of an extensive tract, including much of the coast region of New Granada, lying only a few degrees from the equator, and rich in all sorts of tropical produce, I had formed a very undue idea of its importance. Although the rise and fall of the tide are very moderate on this coast, the ricketty wooden wharf could not be reached at low water. There was nothing for it but to land on the mud, and scramble up the slippery slope to the top of the bank of half-consolidated marl, from twenty to forty feet above the shore, on which the little town is built. It consists of some two hundred houses and stores, nearly all mere plank sheds, but, as usual throughout South America, the inhabitants rejoice in dreams of future wealth and importance to be secured by a railway communicating with the interior. There was no time to be lost; notice had been given that the ship's stay was to be very brief, and even before landing it was apparent that the tropical forest was close at hand. In truth, the last houses are within a stone's throw of the skirts of the forest. Just at this point I was attracted by a leafless bush, evidently one of the spinous species of *Solanum*, with large, yellow, obversely pear-shaped fruits. As

I was about cutting off a specimen, the people, who here seemed very friendly, rushed out of the nearest house and vociferated in warning tones, "Mata! mata!" I was afterwards assured that the fruit is here considered a deadly poison. It appears to be one of the rather numerous varieties of *Solanum mammosum*, a species widely spread through the hotter parts of America.

Being warned not to go out of hearing of the steam-whistle that was to summon us back to the ship, I was obliged to content myself with three short inroads into the forest, through which numerous paths had been cleared. The first effect was perfectly bewildering. The variety of new forms of vegetation surrounding one on every side was simply distracting. Of the larger trees I could, indeed, make out nothing, but the smaller trees and shrubs, crowded together wherever they could reach the daylight, were more than enough to occupy the too short moments.

Of the general character of the climate there could be no doubt. In spite of the blazing sun, with a shade temperature of about 85° Fahr., the ground was everywhere moist. Ferns and *Selaginellæ* met the eye at every turn, with numerous *Cyperaceæ*; and in an open spot, among a crowd of less familiar forms, I found a minute *Utricularia*, scarcely an inch in height. But the predominant feature, and that which interested me most keenly, was the abundance and variety of *Melastomaceæ*. Within the first ten minutes I had gathered specimens of seven species, all of them but one large shrubs. Of the climbers and parasites that give its most distinctive features to the tropical forest, I could in so hurried a peep make out very little. I owe one beautiful species, hitherto undescribed, to my friend W——, who, having wandered in another direction, spied the scarlet flowers of the epiphyte, which I have named *Anthopterus Wardii*, on the trunk of a tree, which was promptly climbed by the active negro who had accompanied him.*

Too soon came the summons of the steam-whistle. As we called on our way at the office of the Pacific Company's agent, we were shown a number of the finer sort of so-called Panama hats, which are chiefly made on this part of the coast. Even on the spot they are expensive articles, a hundred dollars not being considered an unreasonable price for one of the better sort.

Some writers of high authority on geographical botany have held that the most marked division of the flora of tropical South America is that between the regions lying east and west of the Andes. It would be the extreme of rashness for one who has seen so little as I have done of the vegetation of a few scattered points in so vast a region to attempt to draw conclusions from his own observations; but, on the other hand, writers in Europe, even though so learned and so careful as Grisebach and Engler, are under the great disadvantage that the materials available, whether in botanical works or in herbaria, are generally incomplete as regards localities. How is it possible to form any clear picture of the flora of a special district when so large a proportion of the

---

* For a list of the plants collected here, see a paper in the *Journal of the Linnean Society*, vol. xxii.

plants recorded are merely said to come from "Columbia" or "Ecuador," the one larger than Spain, France, and the Low Countries put together, the other equal in extent to the Austrian Empire, and both traversed by mountain ranges varying from fifteen thousand to over eighteen thousand feet in height? I shall have later to make some remarks on the climatal conditions of the coast region extending from Panama to the Bay of Guayaquil, but I may here mention that when I afterwards acquired some slight acquaintance with the flora of Brazil, I was struck with the fact that, although separated by an interval of nearly three thousand miles, and by the great barrier of the Andes, the plants seen in and around the forest at Buenaventura were almost all nearly allied to Brazilian forms.

Further reflection, and such incomplete knowledge as I have been able to acquire as to the flora of inter-tropical South America, lead me to the conclusion that the present vegetable population of this vast region is, when we exclude from view a certain number of immigrants from other regions, mainly derived from two sources. There is, in the first place, the ancient flora of Guiana and tropical Brazil, which has gradually extended itself through Venezuela and Columbia, and along the Pacific coast as far as Ecuador, and, in an opposite direction, through Southern Brazil, to the upper basins of the Uruguay, the Paraná, and the Paraguay. The long period of time occupied by the gradual diffusion of this flora is shown by the large number of peculiar species, and not a few endemic genera that have been developed throughout different parts of this vast region, whose nearest allies, however, are to be found in the original home, Guiana or Brazil. Along with this stock, which mainly occupies the lower country, we find, especially in Venezuela, Columbia, and Ecuador, the modified descendants of vegetable types characteristic of the Andes. Of the Andean flora I shall have something to say in a future page; but I may express the belief that if we go back to the remote period when most of the characteristic types of the vegetation of South America came into existence, we must seek the ancestors of the Brazilian flora, and to a large extent also those of the Andean flora, in the ancient high mountain ranges of Brazil, where we now see, in the vast extent of arenaceous rocks, and in the surviving pinnacles of granite, the ruins of one of the greatest mountain regions of the earth.

Early on Easter Sunday morning, April 9, we were off Tumaco, a small place on one of a group of flat islands lying at the northern extremity of the coast of Ecuador.* These islands are of good repute as having the healthiest climate on this coast. Although close to the equator, cattle are said to thrive, and, if one could forget the presence of a fringe of cocos palms along the shore, the island opposite to us, in great part cleared of forest, with spreading lawns of green pasture, might have been taken for a gentleman's park on some flat part of the English coast. We here parted with General Prado, ex-

* Much cinchona bark, coming from the interior, was formerly shipped at Tumaco; but between horrible roads and the reckless waste of the forests through mismanagement, but little is now conveyed by this way.

AN EX-PRESIDENT OF PERU.

president of Peru, who has purchased one of the islands, and hopes to end his days peacefully as a cattle-breeder. Nothing in his manner or conversation announced either energy or intelligence, but it is impossible not to recognize some kind of ability in a man who, having held such a post at such a time, not only succeeded in escaping the ordinary fate of a Peruvian president—his two immediate predecessors having been assassinated—but also in snatching from the ruin of his country the means of securing an ample provision for himself at a safe distance from home.

In the almost cloudless weather that has prevailed for some days, the apparent path of the sun could not fail to attract attention. Being still so near the vernal equinox, this could not be distinguished from a straight line. Rising out of the horizon at six o'clock, the sun passed exactly through the zenith, and went down perpendicularly in the west into the boundless ocean. Who can wonder that this daily disappearance of the sun has had so large a share in the poetry and the religion of our race? In every land, under every climate, it is the one spectacle which is ever new and ever fascinating. Use cannot stale it; and knowledge, which is said to be driving the imagination out of the field of our modern life, has done nothing to weaken the spell.

We awoke next day to find ourselves in the southern hemisphere, having crossed the line about three a.m. As the morning wore on we passed abreast of the Cabo San Lorenzo, and towards evening, keeping nearer to the coast, were within a few miles of

Cabo Santa Elena. This forms the north-western headland of the Gulf of Guayaquil, a wide bay that extends fully a hundred miles eastward from the coast-line.

At daybreak, April 11, we were inside the large island of Puna, and soon after entered the mouth of the river Guayas. Although it drains but a small district, this has a deep channel, as wide as the Thames at Gravesend, making the town of Guayaquil, which is about thirty miles from its mouth, the natural port for Western Equatorial America. As we steamed northward up the stream, every eye was turned eastward with the hope of descrying some part of the chain of the Andes. It was, indeed, obvious that a great mountain barrier lay in that direction, and beneath the eastern sun dark masses from time to time stood out to view; but along the crest of the range heavy banks of cloud constantly rested, and the summits remained concealed. We knew that the peak of Chimborazo is scarcely more than seventy miles distant from Guayaquil, and is easily seen from the town in clear weather; but we did not know that clear weather is a phenomenon that recurs only on about half a dozen days in the course of the year, and it is needless to say that we did not draw one of these prizes in the lottery. I had been conscious of a distinct change of climate during the preceding night, and this was still more marked after we entered the river. The increase of temperature was but trifling. The thermometer at sea during the two preceding days had ranged from 77° to 79°, and here at nine a.m. it marked only 80°; nor did it ever rise above 84°

while we lay opposite Guayaquil. But the sense of oppressive closeness was more or less felt by every one, and, whatever may be the cause, it seems safe to conclude that the notoriety of this city as one of the most unhealthy in South America is intimately connected with it.

There is, no doubt, much yet to be learned as to the effects of climate on the human constitution, but a few points seem to be sufficiently ascertained. To those whose constitution has been hereditarily adapted to a temperate or cold climate, the enfeebling effect of hot countries depends much more on the constant continuance of a high temperature than on its amount. A place with a mean temperature of 80° Fahr., which varies little above or below that point, is far more injurious to a European than one where intervals of great heat alternate with periods of cooler weather. Still more important, perhaps, is the effect of a hot climate in places where the air is habitually nearly saturated with aqueous vapour. When the temperature of the skin is not much greater than that of the surrounding air, if this be near the point of saturation but little evaporation can take place from the surface. The action of the absorbent vessels is thus checked, and the activity of all the functions is consequently lowered. As it usually happens that the two agencies here discussed act together in tropical countries, the places having a uniform temperature being also for the most part those having an atmosphere heavily charged with vapour, it is easy to understand that Europeans whose vitality is already depressed are especially exposed to suffer from whatever causes induce endemic or epidemic disease. The difficulty in connection with this subject is to explain certain exceptions to the general rule. In several places in the tropics, usually insular stations, where a steady high temperature is combined with the presence of much vapour, the climate is said to have no injurious effects. But the most marked exception seems to be that of seamen. Excluding that large majority whose calling involves frequent changes of climate, there must be now a considerable body of experience respecting those who for a series of years have navigated tropical seas exposed to nearly uniform temperature. I am not aware that there are any facts to sustain the supposition, which might *à priori* seem plausible, that such a life tends to enfeeble the European constitution.

Between a broad fringe of mangrove swamp, backed by a narrow border of forest on either bank, with little to break the monotony of the way, we reached Guayaquil before ten a.m. Seen from the river, with many large buildings and stores covering more than a mile of frontage on the western bank, and a straggling suburb stretching to the base of a low hill to the northward, the city presents an unexpectedly imposing appearance. The present amount of trade is inconsiderable, but if ever these regions can attain to the elementary conditions of good government the development of their natural resources must entail a vast increase of business. The territory of Ecuador includes every variety of climate, and is in great part thoroughly suited to Europeans. All tropical products are obtainable, and, with good management and kindly treatment, the supply of efficient negro labour at moderate wages is considerable. Among other products of the soil, the tobacco of the country about Guayaquil deserves to be better known. Of the many varieties of the coarser kind which are grown throughout Central and South America, this appears to me the best, as it certainly is the cheapest. The hawkers who came on board sold at less than seven shillings a hundred cigars of very fair quality, making, as I was told, a profit of fifty per cent.

It might be not unworthy of the notice of the great steamboat companies to recommend to their agents some little consideration for passengers who travel to see the world. It commonly happens that on the arrival of a steamer, after the first conference between the agent and the captain, a time is fixed for departure which has no relation to the hour really intended. We were told this morning that the steamer was to start at one p.m. The time was clearly too short for an excursion to the neighbouring country, and the inducement to spend a couple of hours in the streets of such an unhealthy town was very trifling. Two young Englishmen went up the river in a boat with the hope of shooting alligators. These creatures abound along the banks of the Guayas, basking in the mud, and looking from a distance like the logs that are floated down by the stream. Our sportsmen had the usual measure of success, and no more. For a bullet to pierce the dense covering that shields this animal is a happy accident, but it suffices to disturb the creature from his rest, and to induce him to crawl or roll into the river, and to accomplish this is at least a new experience. Through the courtesy of a native gentleman, the travellers were induced to land at a *hacienda* on the river, where horses were provided, and they galloped back to the town before one o'clock. Meanwhile the Jamaica story was repeated. It was announced that the agent had decided to keep the steamer till three p.m.; and finally we learned that we should remain at our moorings till early next morning.

On her last voyage the *Islay* had started too late; night fell before she cleared the mouth of the river, and, in the dark, she had run down a *chatta*—one of the cumbrous native barges that ply along the stream. Of fifteen natives in the barge thirteen were saved, three of them by the courage and activity of the chief officer, who jumped into the river to their rescue. Our captain very properly objected to the risk of another similar accident, and decided to wait for daylight. The cause of the delay remained a mystery, for all that was shipped of passengers and cargo was of a kind that did not seem likely to be very remunerative. At first sight it appeared merely as a characteristic of a rude state of society that the country people around Guayaquil are used to embark on the southward-bound steamers with tropical fruit raised by themselves, which they carry to Lima, and even as far as Valparaiso, dispose of at a handsome profit, and then return home. As most of the profit must go into the coffers of the Pacific Steam Company, the motive is not very obvious; but after a little further experience I fully understood it. Even if they clear little more than the price of their passage, these people find their advantage in undertaking an annual expedition of this kind. Apart from the very positive benefit to health, they gain what they like most in the world—a season of absolute idleness, with the amusement of seeing new objects and talking to new people. For the remainder of the voyage the main-deck was crowded and somewhat encumbered by picturesque groups of rough men, some accompanied by womankind, alternating with huge heaps of tropical fruit—pineapples and bananas, a single bunch of the latter sometimes weighing more than a hundred pounds.

The thermometer scarcely varied by a small fraction from 80° throughout the night and the following day, until we had cleared the Gulf of Guayaquil; and even at this moderate temperature the feeling of lassitude continued as on the previous day. Of the famous mosquitos of the river Guayas we had little experience. They are said sometimes to attack in swarms so numerous and ferocious that, even by day, it becomes difficult for officers and men to manage a ship on the river.

The sun had set on the following evening, April 12, before we were well abreast of Cabo Blanco, the southern headland of the Gulf of Guayaquil, and we saw nothing of its southern shore. About one-half of this belongs to Peru, and close to the frontier-line is the little port of Tumbez, sometimes visited by passing steamers. I was assured by two of the ship's officers that the climate and vegetation of this place are much the same as at Guayaquil, but there are few parts of the American coast that better deserve careful examination by a scientific naturalist.

During the night of the 12th we passed Cape Parinas, the westernmost headland of South America, and before sunrise were in the roads of Payta. Being aware that the so-called rainless zone of Peru extends northward to this place, I was especially anxious to see as much of it as possible. During the night the temperature had fallen, especially after rounding Cape Parinas, and at sunrise stood at 74°. In the cooler air, and under the excitement of pleasant anticipation, the lassitude of the two preceding days utterly disappeared; and as day dawned I stood on deck, with my tin box slung to my back, ready to go ashore long before there was any possibility of doing so. The officers told me, indeed, that there was no use in taking a botanical box, as the country about Payta was absolutely without vegetation. I have many times had the same assurance given me, but the time had not yet come when I was to find it correct, and I felt that Payta was not one of such rare spots on the earth.

The appearance of the place and of its surroundings is unquestionably very strange, and the contrast between it and the shores of the neighbouring Gulf of Guayaquil is simply marvellous. Saving the presence of a mean little modern church, with two shabby wooden towers coated with plaster, the aspect of the little town reminded me of Suez, with the difference that the surrounding desert is here raised about a hundred feet above the sea-level. The place, I presume, is improved since it was visited and described by Squiers, and I found that on the slope between the base of the plateau and the beach there is ample space for some mean streets.

With several companions who were kind enough to interest themselves in plant-hunting, I at once turned towards the sea-beach at the south-western side of the town, keeping along the base of the low cliffs that here descend to the water's edge. The seaward face of the cliffs is furrowed by numerous gullies, and in one of the broadest of these I was delighted to observe numerous stunted bushes well laden with crimson flowers. This turned out to be *Galvesia limensis*, a plant found only at a few spots in Peru, whose nearest but yet distant European ally is the common snapdragon. In the upper part of the same gully were the withered remains of several other species, most of which have been since identified. Emerging on the plateau, we found ourselves on a wide plain, apparently unbroken, leading up to a range of hills some fifteen or twenty miles distant. Though we were here only five degrees from the equator, and before we returned to the ship the sun had risen as high as on a summer's noon in England, the southerly breeze felt delightfully cool and fresh, and at midday, under the vertical sun, the temperature on board ship was not quite $75°$.

Vegetation, as I anticipated, was not entirely absent from the plateau, but it was more scarce than I had anywhere seen it, except in the tracts west of the Nile above Cairo, where the drifting sands covers up and bury everything on the surface. In the northern Sahara, about Biskra, where rain is much less infrequent than here, vegetation, though scanty, is nearly continuous, and it is not easy to find spaces of several square yards absolutely without a single plant. About Suez, and on parts of the isthmus where a slight infiltration from the sweet water canal has not developed a more varied vegetation, the number of species in a given tract is often very limited; but tufts of vigorous growth, especially of the salt-loving species, are seen at frequent intervals. On the plateau of Payta, where, as we rambled about, several pairs of eyes were on the alert, but a single tuft of verdure visible at a distance could be made out. This was formed by several bushes of *Prosopis limensis* growing together. Elsewhere the few plants seen were confined to the occasional shallow depressions where rain rests longest. All, of course, had perennial roots, and scarcely one of them rose as much as three inches from the ground.*

I found it difficult to account for the origin of the sands which are sparingly scattered over the plateau, but accumulated to a considerable depth on the slopes behind the town. The underlying rock seen in ascending to the plateau is a tolerably compact shale; but the hard crust forming the superficial stratum appears to consist of different materials, and not to be made up from the disintegrated materials of the shale. At several places, both below the cliffs and on the plateau, I found large scattered fragments of what appeared to be a very recent calcareous formation, largely composed of shells of living species; but this was nowhere seen *in situ*, and I was unable to conjecture the origin of these fragments.

* For a list of the species collected, see the *Journal of Linnean Society*, vol. xxii.

Before returning to the *Islay*, I had the advantage of a short conversation with the very intelligent gentleman who acts as British consular agent at Payta, and whose ability would perhaps be seen to advantage in a more conspicuous post. The information received from him fully confirmed the impressions formed during my short excursion. The appearance of the gullies that furrow the seaward face of the plateau sufficiently showed that, however infrequent they may be, heavy rains must sometimes visit this part of the coast. I now learned that, in point of fact, abundant rain lasting for several days recurs at intervals of three or four years, the last having been seen in the year 1879. As happens everywhere else in the arid coast zone, extending nearly two thousand miles from Payta to Coquimbo in North Chili, abundant rainfall is speedily followed by an outburst of herbaceous vegetation covering the surfaces that have so long been bare. During the long dry intervals slight showers occur occasionally a few times in each year. These are quite insufficient to cause any general appearance of fresh vegetation, but suffice, it would seem, to maintain the vitality of the few species that hold their ground persistently. The ordinary supply of water in Payta, obtained from a stream descending from the Andes seventeen miles distant, is carried by donkeys that are despatched every morning for the purpose. There was something quite strange in the appearance of a few bundles of fresh grass which we saw in the *plaza*. They had come that morning by the same conveyance for the support of the very few domestic animals that it is possible to keep in such a place.

The problems suggested by the singular climatal conditions of this region of South America have not, I think, been as fully discussed as they deserve to be, and I here venture on some remarks as a contribution to the subject.

The existence of the so-called rainless zone on the west coast of South America is usually accounted for by two agencies whose union is necessary to produce the result. The great range of the Andes, it is said, acts as a condenser on the moisture that is constantly carried from the Atlantic coasts by the general westward drift of the atmosphere in low latitudes. The copious rainfall thus produced on the eastern slopes of the great range leaves the air of the highlands of Peru and Bolivia relatively dry and cool, so that any portion that may descend to the coast on the western declivity tends to prevent rather than to cause fresh aqueous precipitations. Meanwhile the branch of the Antarctic Ocean current known as the Humboldt current, which sets northward along the sea-board from Western Patagonia, is accompanied by an aërial current, or prevailing breeze, which keeps the same direction. The cold air flowing towards the equator, being gradually warmed, has its capacity for holding vapour in suspension constantly increased, and is thus enabled to absorb a large portion of the vapour contained in the currents that occasionally flow inland from the Pacific, so that the production of rain is a rare event, recurring only at long intervals. Admitting the plausibility of this explanation, a first difficulty presents itself. If the Andes act as a barrier against the vapour-laden atmosphere of eastern tropical America throughout Peru, Bolivia, and Northern Chili why, it may be asked, do they fail to perform the same function in Ecuador and Colombia? Whence the absolute contrast in point of climate that exists between these regions? Why is the littoral zone between the Gulf of Guayaquil and that of Panama, a distance of some eight hundred miles, not merely less dry than that of Peru, but actually more moist that most parts of the coast of Brazil or Guiana?

Some answer may, I think, be given to these questions. In the first place, comparing the orography of Peru and Bolivia with that of Ecuador, some important differences must be noted. In Eastern Peru, as is at once shown by the direction of the principal rivers, we find no less than four parallel mountain ranges, increasing in mean elevation as we travel from east to west. The westernmost range, to which in Peru the name *Cordillera* is exclusively applied, does not everywhere include the highest peaks, but has the highest mean elevation. The second range, exclusively called *Andes* in Peru, rivals the first in height and importance. I know of no collective names by which to distinguish the third range, dividing the valley of the Huallaga from that of the Ucayali, nor the fourth range, forming the eastern boundary of the latter stream. In South Peru and Bolivia the mountain ranges are less regularly disposed, but cover a still wider area; and throughout the whole region it is obvious that the warm and moist currents drifting slowly westward have to traverse a zone of lofty mountains varying from four to six hundred miles in width, and can carry no moisture available to produce rain on the western seaboard. In Ecuador the two principal ranges—the Cordillera and the Andes—are much nearer together than they usually are in Peru, and no parallel ranges flank them on the east. The numerous tributaries of the Maranon flow in a tolerably direct course east or south-east, many of them rising within a hundred and fifty miles of the Pacific coast. It follows that the atmospheric currents meeting less preliminary obstruction reach the eastern slopes of the main range still very heavily charged with vapour. In crossing the barrier a large portion of the burthen must be deposited; but it is probable that a large amount is nevertheless carried to the western side of the range.

It may be said that this explanation, whatever it may be worth, cannot apply to the territory of Colombia, where the Andes are broken up into at least three lofty ranges, and the mountains cover as wide a space as they do in Peru. My impression is that the abundant supply of moisture on the west coast of Colombia arises from a different source. The effects of the Isthmus of Panama as a barrier against atmospheric currents must be absolutely insignificant, and I have no doubt that those which flow eastward along the coast of the Caribbean Sea are in part diverted south-east and south along the west coast of Colombia.

There can, however, be little doubt that in determining the climate of the west coast the influence of the Humboldt current, and of the cool southerly breezes that accompany it, is far greater than that of the disposition of the mountain ranges. A glance at the map shows that about the fifth and sixth degrees of south latitude the direction of the coast undergoes a considerable change. On the voyage from Panama, we had hitherto steered somewhat west of south; henceforward our course lay between south-south-east and south-east. All the currents of the ocean and atmosphere, whose existence arises from the unequal distribution of heat on the earth's surface, vary somewhat in their course throughout the year with the changes of season, and this doubtless holds good on the American coast. I believe, however, that both the sea and air currents from the south are normally deflected away from the coast at the promontory of Ajulla (sometimes written "Ahuja"), a short distance south of Payta. A further portion is again deflected westward at Cape Parinas, north of which headland they seem not to be ordinarily met. I infer, however, from the testimony of seamen, that at some seasons they are felt near the coast as far north as the equator, and even beyond it. This inference was confirmed by observing the parched appearance of the seaward slope of Cabo Sta. Elena, north of the Gulf of Guayaquil, which apparently does not fully share in the frequent rains that elsewhere visit the coast of Ecuador.

Whatever force there may be in the above suggestions, I confess that they do not seem to me adequate to account for the extraordinary difference of climate between places so near as Payta and Tumbez—not quite a hundred miles apart—and I trust that further light may be thrown upon the matter by a scientific traveller able to spare the necessary time. So far as I know, no such abrupt and complete a change is known elsewhere in the world. I was unable to obtain any information as to a range of hills or mountains, marked in Arrowsmith's map "Sa. Amatapi," which appears to extend east or east-north-east from Cape Parinas. Its height can scarcely be considerable, as it does not appear to have attracted the attention of the seamen who are familiar with this coast; but, on the other hand, there is some reason to think that the southerly breezes prevailing on the coast do not extend to any great height above the sea-level. It would be interesting if we should find on the opposite sides of a range of unimportant hills the same contrasts of climate and vegetation that are known to prevail between the eastern and western slopes of the Peruvian Andes.*

Along the coast of Northern Peru are numerous small islets, evidently at some period detached from

---

* The abrupt change in the vegetation on this part of the American coast has been noticed by Humboldt, Weddell, and other scientific travellers. In a note to the French edition of Grisebach (" Vegetation du Globe," traduit par P. de Tchihatcheff, ii. p. 615), M. André expresses the opinion that this, as well as some other cases of abrupt change in the vegetation observed by him in Colombia, are to be explained by the nature of the soil, which in the arid tracts is sandy or stony, and fails to retain moisture. Admitting that in certain cases this may afford a partial explanation of the facts, it is scarcely conceivable that the limit of the zone wherein little or no rain falls should exactly coincide with a change in the constitution of the soil, and I should be more disposed to admit a reversed order of causation, the porous and mobile superficial crust remaining in those tracts where, owing to deficient rainfall, there is no formation of vegetable mould, and no accumulation of the finer sediment forming a retentive clay.

the continent either by subsidence or by marine erosion. Here, in the almost complete absence of rain, were formed those secular accumulations deposited by sea-birds, which, when known in Europe under the name of *guano*, suddenly rivalled the mines of the precious metals as sources of easily acquired wealth. The two most considerable groups are respectively named Lobos de tierra and Lobos de afuera; a smaller group near to Payta is also called Lobos. At the western end of the largest of the latter group the waves have excavated a natural arch, which, after a sufficient period of further excavation, will fall and give rise to a new detached islet. A brisk southerly breeze made the air feel cooler than it had done since we entered the tropics, as we ran about due south until sunset, when, after passing abreast of the promontory of Ajulla, our course was altered to nearly due south-east. I was assured by a native passenger that the promontory of Ajulla, for a distance of thirty or forty miles, is an absolute desert, without a drop of water or the slightest trace of vegetation. Experience has made me somewhat sceptical as to statements of this nature made by non-scientific observers. During the day we frequently observed a fish which appears distinct from the flying-fish of the Atlantic. The pectoral fins appear to be less developed, and in consequence the flight is shorter, and the animal seems to have less command over its movements.

Our course on April 14 lay rather far from land. It was known that yellow fever had broken out at Truxillo, and it was decided that we should run direct to Callao, without touching at that or any of the smaller places on the coast sometimes visited by the steamers. Although the air appeared to be somewhat hazy, the range of the Cordillera, more than a hundred miles distant, was distinctly seen in the afternoon. Very soon after we ran into a dense bank of fog, in which we were immersed for several hours, our cautious captain remaining meanwhile on the bridge, and the frequent cry of the steam-whistle ceased only when we steamed out of the fog into a brilliant star-lit night.

These fogs, which are frequent along the Peruvian coast, are the chief, if not the only, difficulty with which the navigator has to contend. When they rest over the land it becomes extremely difficult to make the ports, and at sea they involve the possible risk of collision. If this risk is at present but slight, it must become more serious when intercourse increases, as it must inevitably do if the Ship Canal should ever be completed; and for the general safety it may be expedient to prescribe special rules as to the course to be taken by vessels proceeding north or south along the coast. The origin of the fogs must be obvious to any one who considers the physical conditions of this region, to which I have already referred. The air must be very frequently near the point of saturation, and a slight fall of temperature, or the local intermixture of a body of moister air, must suffice to produce fog. The remarkable thing is that this should so very rarely undergo the further change requisite to cause rain. To some young Englishmen on board, the remarkable coolness of the air along this coast was a continual subject of jesting comment; and on more than one occasion the "Tropics" were emphatically declared to be "humbugs." It is certain that for thirty-six hours before reaching Callao the shaded thermometer never reached 70°, and stood at noon, with a clear sky and a brisk southerly breeze, no higher than 68°.

CHAPTER II.

Arrival at Callao—Quarantine—The war between Chili and Peru—Aspect of Lima—General Lynch—Andean railway to Chicla—Valley of the Rimac—Puente Infernillo—Chicla—Mountain-sickness—Flora of the Temperate zone of the Andes—Excursion to the higher region—Climate of the Cordillera—Remarks on the Andean flora—Return to Lima—Visit to a sugar-plantation—Condition of Peru—Prospect of anarchy

THE steam-whistle, sounding about daybreak on April 15, announced that we were again wrapped in fog. As the *Islay* advanced at half speed the fog lightened without clearing, until about nine a.m. we made the island of San Lorenzo, and, as the haze finally melted away into bright sunshine, found ourselves half an hour later in the harbour of Callao. The moment was exciting for those who, like myself, approached as strangers the shore which had in our childhood seemed so strange, so adventure-fraught, so distant. Already some one had pointed out the towers of the Cathedral of Lima, with the Cordillera apparently so near that the mountains must begin outside the gates. All stood on deck prepared to land—some already looking forward to luncheon in the city of Pizarro—and waiting only for the usual formalities of the visit of the *sanidad*. At length the officials came, and, after the usual parley over the ship's side, it became apparent that the visit was no mere formality. At last the ominous word *quarantine* was heard, received at first with mere incredulity, as something too absurd, but at last taking the consistence of a stern fact. Since the outbreak of yellow fever among the troops at Truxillo, the Chilian authorities have naturally become nervously anxious to protect the occupying army from this danger, and every precaution is put in force. Under these circumstances, a ship coming from Guayaquil was naturally an object of suspicion. There certainly was not at the time any epidemic fever at that place; but, if reports be true, sporadic cases are not unfrequent, and that city is rarely, if ever, quite free from malignant zymotic disease. At last the discussion was closed, by a definite order that we should repair to the quarantine ground under the lee of the island of San Lorenzo.

Up to this time we had scarcely given attention to the scene immediately surrounding us; yet the harbour of Callao is at any time an interesting sight, and at this moment its aspect was peculiarly expressive. Although the Chilian forces had before this time become absolute masters of the entire seaboard of Peru, and there was no reason to apprehend any renewal of the struggle by sea, the memorials of the desperate encounters which marked the earlier phase of the war were here still fresh. Near the shore in several different directions were the wrecks of ships which had sunk while the captors were endeavouring to bring them into harbour, the masts sticking up idly above water and doing the duty of buoys. Still afloat, though looking terribly battered and scarcely seaworthy, was that remarkable little ship, the *Huascar*, looking a mere pigmy beside the warships in the harbour from which the Chilian, American, French, and Italian flags were flying, England being for the moment unrepresented.

The naval war between Chili and Peru was conducted at such a distance from Europe, and its causes were so little understood, that it excited but feeble interest. Even the circumstance that, in an encounter brought about by the incompetence and rashness of a British commander, the pigmy Peruvian force was able with impunity to inflict an affront on the national flag, scarcely excited in England more than momentary surprise. Nevertheless the story of the war, which yet awaits an impartial chronicler,* abounds with dramatic incident. The record is ennobled by acts of heroic bravery on both sides, while at the same time it suggests matter for serious consideration to the professional seaman. The important part which small fast ships, carrying one or two heavy guns only, may play in the altered conditions of naval warfare has been often pointed out, but has been practically illustrated only in the war between Chili and Peru. It does not seem as if the importance of the lesson had been yet fully appreciated by those responsible for the naval administration of the great European powers.

For the remainder of the day, and during the whole

---

* The only detailed account of the operations that I have seen is in a work entitled, "Histoire de la Guerre du Pacifique," by Don Diego Barros Arana. Paris: 1881. It appears to be fairly accurate as to facts, but coloured by very decided Chilian sympathies.

of the 16th, we lay at anchor about half a mile from the shore of the island of San Lorenzo, a bare rough hill, mainly formed, it would seem, of volcanic rock overlaid in places by beds of very modern formation. All naturalists are familiar with the evidence adduced by Darwin, proving the considerable elevation of the island and the adjacent mainland since the period of the Incas, as well as Tschudi's arguments going to show that in more recent times there has been a period of subsidence.

Of the objects near at hand the most interesting were the large black pelicans which in great numbers frequent the bay or harbour of Callao, attracted, no doubt, by the offal abundantly supplied from the town and the shipping. Seemingly indefatigable and insatiable, these birds continued for hours to circle in long sweeping curves over the water, swooping down on any object that attracted their appetite. The body appears to be somewhat slighter than that of the white pelican of the East, but the breadth of wing and length of the neck are about the same. When on the wing the plumage appears to be black, but in truth it is of a dark bluish slate colour.

Our detention in quarantine might have been prolonged but for the fortunate circumstance that the contents of the mail-bags carried by the *Islay* were at this moment the object of anxious curiosity to the Chilian authorities, and to the representatives of foreign powers. The position of affairs was already sufficiently critical, and the attitude recently assumed by the Government of the United States had added a new element of uncertainty to the existing difficulties.

Mr. Hurlbut, the last American representative, had died, and Mr. Trescott, who supplied his place, was ostensibly charged with the attempt to bring about a peace between Chili and Peru, but was supposed to be chiefly intent on extricating his Government from a position into which it had been led by a series of proceedings which had neither raised the national reputation nor secured the good-will of either Chili or Peru.

While we lay off the harbour, watched day and night by the crew of a launch stationed beside us to prevent communication with the land, we received three successive visits from the officers of the American man-of-war lying in the harbour, who approached near enough to hold conversation with our captain. The message was a request, finally conveyed in somewhat imperious terms, that the despatches addressed to the American envoy should at once be delivered. The American foreign office is not, I believe, accustomed to forward diplomatic despatches in a separate bag, but merely uses the ordinary post. Our captain properly declined to take the responsibility of opening the mail-bags, which he was bound to deliver intact to the postal authorities as soon as we were admitted to pratique. The result was that on Monday, just as we were beginning to be seriously uneasy at the prospect of a long detention, a steam launch was seen to approach, having a number of officials on board. A seemingly interminable conversation between these and the captain and medical officer of our ship finally resulted in a Chilian medical man coming on board to make a careful examination of the ship, the crew, and the passengers. After we had been duly marshalled and inspected—the first-class passengers on the spar-deck, the others on the main-deck—the welcome announcement, "Admitted to pratique," ran through the ship. Not much time was lost in moving up to the proper moorings in the harbour, some two miles distant, and about noon we were set on land close to the custom-house.

The boatmen, the porters, and the nondescript hangers-on about the quays of a port, formed a strange and motley assemblage, in whose countenances three very distinct types of humanity—the European, the negro, and the South American Indian—were mingled in the most varied proportions, scarcely one denoting an unmixed origin. The arrangements at Callao are convenient for strangers. The customhouse officers, though unbribed, gave no trouble, and the rather voluminous luggage of six English passengers was entrusted to a man who undertook for ten *soles* (about thirty-three shillings) to convey the whole to the chief hotel in Lima. No time was left to see anything of Callao. A train was about to start; and in half an hour we were carried over the level space—about seven and a half miles—that separates Lima from the port of Callao.

Occupied by the forces of her victorious rival, and shorn of most of the almost fabulous wealth that once enriched her inhabitants, Peru can, even in her present ruined state, show a capital city that impresses the stranger. It is true that the buildings have no architectural merit, that most of the streets are horribly ill-paved, and that at present there is little outward appearance of wealth in the thoroughfares; in spite of all this the general aspect is novel and pleasing. Although violent earthquakes have rarely occurred in this region, slight shocks are very frequent, and remind the inhabitants that formidable telluric forces are slumbering close at hand. Hence, as a rule, the houses have only a single floor above the ground, and cover a proportionately large space. As in Southern Spain, all those of the better class enclose a *patio*, or courtyard, partly occupied by tropical trees or flowering shrubs. Fronting the street, or the *plaza*, a long projecting balcony, enclosed with glass, enables the inmates to enjoy that refuge from absolute vacancy which is afforded by gazing at the passers-by, and which seems to supply the place of occupation to much of the population even in Southern Europe.

With scarcely an exception, the numerous churches are vile examples of debased renaissance architecture, fronted with stucco ornamentation in great part fallen to decay. Not long before our arrival, I believe under the Chilian administration, they had been all freshly covered with whitewash, cut into rectangular spaces by broad bands of bright blue. In the streets near the great *plaza* there was much apparent animation during the day; but the shops were closed an hour before nightfall, and after dark the city was hushed into unnatural silence. The fair *Limeñas*, as to whose charms travellers have been eloquent, and who used to throng the public drives and walks towards sunset, were no longer to be seen. To exhibit themselves would be to display indifference to the misfortunes of their country. Some might be observed, indeed, during the morning hours, plainly dressed in black, going either to church or on some business errand ; but they were so closely wrapped up in a *manta* as to be completely disguised.

On landing in Peru, the one question which completely engrossed my mind was whether or not it would be possible for me, in the present state of the country, to reach the upper region of the Andes.

To a naturalist this great chain must ever be the dominant feature of the South American continent. To its structure and its flora and fauna are attached questions of overwhelming importance to the past history of our planet, and, however little a man may hope to effect during a flying visit, the desire to gain that degree of acquaintance which actual observation alone can give becomes painfully intense. I was aware that what had formerly been a long and rather laborious journey had of late years been reduced to a mere excursion by the construction of two lines of railway, leading from the sea-coast to the upper region. That which, if free to choose, I should have preferred starts from the coast at Mollendo, and, passing the important town of Arequipa, traverses the crest of the Cordillera, and has its terminus at Puno, on the Lake of Titicaca, in the centre of the plateau which lies between the two main ridges of the Andes. The region surrounding this great lake, which here divides Peru from Bolivia, must offer objects of interest only too numerous and too engrossing for a traveller whose time is counted by days. Although the level of the lake is some 12,800 feet above the sea, the peaks of Sorata rise above its eastern shores to a further height of nearly 10,000 feet; and lake steamers give access to most of the inhabited places on its shores—no slight matter when it is remembered that the lake measures more than a hundred miles in length.

The second line, which, starting from the city of Lima, is carried nearly due east along the valley of the Rimac, was designed to open communication by the most direct route between the capital and the fertile region on the eastern slopes of the Andes—called in Peru the *Montaña*—as well as with the rich silver region of Cerro de Pasco. The crest of the Cordillera, or western ridge of the Andes, is scarcely eighty miles from Lima in a direct line, but the most practicable pass is somewhat higher than the summit of Mont Blanc. The road was to pierce the pass by a tunnel 15,645 feet above the sea-level, and thence to descend to the town of Oroya on the high plateau that divides the two main ridges. As the line was laid out, the distance from Lima to the summit-level was only 97 miles, and that to Oroya 129 miles.

Considered merely as engineering works, these lines, which owe their existence to the enterprise of an American contractor and the skill of the engineers who carried out the undertaking, may fairly be counted among the wonders of the world. The Oroya line, the more difficult of the two, unfortunately remained unfinished. Although the loans contracted in Europe by the Peruvian Government more than sufficed to defray the cost of all the industrial undertakings that they were professedly intended to supply, it is scarcely necessary to say that a large portion disappeared through underground channels, leaving legitimate demands unprovided for. The stipulated instalments due to Mr. Meiggs, the great contractor, remained unpaid, and, in the midst of the difficulties in which he was thus involved, his death put a final stoppage to the works. The line had been completed and opened for a distance of about eighty miles from Lima, as far as the village of Chicla, 12,220 feet above the sea. From that time forward Mr. Meiggs devoted his energies to the boring of the tunnel at the summit, probably under the impression that if that were once finished the Peruvian Government could scarcely fail to provide the funds necessary to complete the line on either side.

I had found it impossible to ascertain before leaving England what had been the fate of these magnificent works since the ravages of war had devastated the region through which they are carried. Various quite inconsistent stories had reached me through the passengers from Panama, Guayaquil, and Payta. Traffic, said some, continued on both lines just as before the war; traffic, said others, had been completely stopped by order of the Chilian authorities; others, finally, asserted that the Oroya line had been so damaged by either belligerent as to be rendered permanently useless.

Before I had been many hours on shore, I was able to get authentic information which relieved my mind from further anxiety. The southern line, from Mollendo to Puno, was open; but Arequipa, the chief place on the way, was still in possession of the Peruvians, who occupied it in some force. With permits, to be

F

obtained from the commanding officers on both sides, it might be possible to go and return, supposing no fresh outbreak of hostile movements of the troops on either side. The news as to the Oroya line was even more satisfactory. The whole line was occupied by the Chilian forces, there being a detachment at Chicla, with outposts on the farther side of the pass. The line had been for some time closed to traffic, but had been re-opened a few days before our arrival. With a permit, to be obtained from the chief of the staff in Lima, there would be no difficulty in proceeding to Chicla.

My decision was speedily taken. Under the most favourable circumstances, the time necessary to reach Puno and return to the coast, with the not improbable risk of detention, was more than I could afford. Further than this, as Puno lies on the plateau remote from the mountains, I should see but little of the characteristic flora of the Andes, unless I could reach some place on the eastern shore of the Lake of Titicaca, whence access could be had to the flanks of the Sorata Andes.

Some description of the Lake of Titicaca which I had read as a boy still dwelt in my mind, and the memoirs and conversation of the late Mr. Pentland had long made the peaks of Sorata objects of especial interest to me. There could, however, be no doubt that the faint hope of beholding them which had lingered till then must be renounced, and I was too happy at the prospect of achieving a short visit to the more accessible part of the chain to have leisure for any keen regret.

Having ascertained that the trains to Chicla departed only every second day, returning thence on the alternate days, I arranged to start on the 20th. During the two intermediate days, I had the opportunity of making several agreeable acquaintances. Sir Spencer St. John, the English minister, had lately returned to Europe, and the legation was temporarily under the charge of Mr. J. R. Graham, who had recently acted as *chargé d'affaires* in Guatemala. Among other kind attentions which I have to acknowledge, Mr. Graham was good enough to introduce me to Don Patricio Lynch, commander-in-chief of the Chilian forces in Peru.

The object of boundless admiration from his own followers, and of still more unmeasured denunciation from his enemies, General Lynch is undoubtedly the most remarkable man who has come to the front during the late unhappy war in South America. Like most of the men who have acquired military renown in that part of the world, he is of Irish extraction, his grandfather having settled in Chili early in the present century. Having served as a young man for a time in the English navy, he was promoted by the Chilian Government, some time after the outbreak of the war, to a naval command. The operations at sea had, up to that time, been on the whole unfavourable to Chili, and the successes which finally changed the aspect of the war by sea were largely ascribed to the energy and ability of Admiral Lynch. Passing from the sea to the land, he so much distinguished himself in various daring encounters with the enemy that he was finally promoted to the chief command of the Chilian forces in Peru, and at this time was virtually dictator, with absolute rule over the whole coast region occupied by the Chilian army.

However open to discussion might be the policy adopted by the Chilians towards the conquered country, there was a general agreement as to one matter of no slight importance. The population of Lima and the surrounding districts is composed of the most varied constituents—native Indian, negro, and the mongrel offspring of the intermixture of these with European blood, to all which of late years has been added a large contingent of Chinese immigrants. It is not surprising that, under inefficient administration, there should have arisen from the dregs of such a population a large class either actually living by crime or ready to resort to outrage as favourable opportunities might arise. On the other hand, the Chilian army, for which there was but a small nucleus of regular troops, had to be largely recruited from among the loose fish of the floating population of South America, and naturally included no small number of bad subjects, ready to make the utmost use of the license of war. For many years past the police of Lima was notoriously inefficient; robberies were frequent, and there were many spots in the neighbourhood of the city where it was considered unsafe to go unarmed even in broad daylight. It was not unreasonably feared that in such conditions the occupation of the city by the Chilians would have results disastrous for the safety of the numerous foreign residents and the peaceful citizens. It was through the energy and capacity of General Lynch that the apprehended reign of disorder was averted. An efficient police was at once established, speedy capital punishment was awarded in every case of serious outrage, and with stern impartiality a short shrift was allotted alike to the Peruvian marauder and the looter wearing Chilian uniform. It was admitted on all hands that the city had never before been so safe, while, at the same time, the ordinary municipal work of cleansing, watering, and lighting the streets and public places had been visibly improved under the stimulus of vigorous administration.

My reception by the Chilian general was all that I could desire. He at once expressed his readiness to assist my objects in every way, and carried out his promise by giving me a letter to the officer commanding the detachment at Chicla, with instructions to provide horses and guides and all needful protection for myself and my companion. I failed to detect in General Lynch any of the characteristics, usually so persistent, of men of Irish descent. The stately courtesy and serious expression, reminding one of the bearing of a Castilian gentleman, were not enlivened by the irrepressible touches of liveliness that involuntarily relieve even a careworn Irishman from the pressure of his environment. One particularity in the arrangements at head-quarters struck me as singular; but I afterwards understood that it was merely the transference to Peru of the ordinary habits of Chili. The head-quarters of the general were fixed in the former palace of the Spanish viceroys. A sentry in the street paid no attention as, in company with Mr. Graham, I entered the first court, and it appeared that every one, or, at least, every decently dressed stranger, was free to pass. Through an open door we entered the first of a suite of large rooms, and advanced from one to another without encountering a human being, whether guard or attendant, until in the last room but one, seemingly by accident, a secretary presented himself, who at once ushered us into the cabinet of the general. In the case of any public man in Europe, to say nothing of the chief of an army of occupation constantly assailed by the fiercest denunciations, and left thus easy of access, some fanatic or madman would speedily translate the popular hatred into grim deed.

Among the acquaintances made in Lima, I must mention the name of Mr. William Nation, a gentleman who, amidst many difficulties, has acquired an extensive knowledge of the fauna and flora of Peru, and has observed with attention many facts of interest connected with the natural history of the country. After my return from Chicla, Mr. Nation was kind enough to accompany me in two short excursions in the neighbourhood of the city, and I am further indebted to him for much valuable assistance and information.

Soon after eight a.m. on the morning of April 20, I started from the railway station at Lima, in company with my friend W——, who was fortunately able to absent himself for some days. The country lying between the coast and the foot of the Cordillera appears to the eye a horizontal plain, but is, in fact, a slope inclining towards the sea, and rising very uniformly about seventy feet per mile.* This ancient

---

* The heights given in the text are those of the railway stations.

sea-bottom extends for a distance of fully fifteen miles from Lima into the valley of the Rimac, which, in approaching the coast, gradually spreads out from a narrow gorge to a wide valley with a flat floor. At the same time the river gradually dwindles from a copious rushing torrent to a meagre stream, running in many shallow channels over a broad stony bed, until it is finally almost lost in the marshes near Callao. Its waters are consumed by the numerous irrigation channels; for it must be remembered that along the western side of the continent, for a distance of nearly thirty degrees of latitude, cultivation is confined to those tracts which can be irrigated by streams from the Andes. Keeping pretty near to the left bank of the Rimac, the railway runs between two detached hills, formerly islands when the sea stood a few hundred feet above its present level. That on the north side is called the Amancais, and another less extensive mass rises south of the river.

Throughout the greater part of the year these hills, as well as the lower slopes of the Cordillera, appear, as they did to me, absolutely bare of vegetation; but in winter, from June to September, slight showers of rain are not unfrequent, and the fogs, denser than in other seasons, rest more constantly on the hills, and doubtless deposit abundant night-dews on the surface. The seeds and bulbs and rhizomes awake from their long sleep, and in a few days the slopes are covered with a brilliant carpet, in which bright flowers of various species follow each other in rapid succession.

Alongside of the railway runs a broad road covered to a depth of a couple of feet with volcanic sand, with occasional loose blocks of stone. The struggles of the few laden animals that we saw in passing, as they toiled along this weary track under a scorching sun, suggested a thought of the wonderful changes which modern inventions have already effected, and are destined to effect in the future, throughout every part of the world. The track before our eyes was, until the other day, the sole line of direct communication between Lima and the interior of Peru. The passage of men and animals had in the course of centuries reduced the original stony surface to a river of fine sand, and by no better mode of transport had the treasures of Cerro de Pasco, and the other rich silver deposits of the same region, been carried to the coast to sap the manhood and energy of the Spanish settlers in Peru, and help to achieve the same result in the mother country.

The American railway car, which is not without its drawbacks for ordinary travellers, is admirably suited to a naturalist in a new country. No time is lost in opening and shutting doors. Standing ready on the platform, one jumps off at every stoppage of the train, and jumps up again without delay or hindrance. I was able to appreciate these advantages during this day, and to add considerably to my collections by turning every moment to account. At first the vegetation was, of course, extremely scanty; but I was interested by finding here some representatives of genera that extend to the hotter and drier parts of the Mediterranean region, such as *Boerhavia* and *Lippia*.

Not far beyond the station of Santa Clara, near to

VEGETATION OF THE RIMAC VALLEY.

which is a large sugar-plantation, the slopes on either side of the valley become more continuous, and gradually approach nearer together. The first trace of vegetation visible from a distance was shown by one of the cactus tribe, probably a *Cereus*, and as we ascended I was able to distinguish two other species of the same family.

At many points in the valley, always on slightly rising ground, shapeless inequalities of the surface marked with their rough outline all that now remains of the numerous villages that in the days of the Incas were scattered at short intervals.

As we advanced, the slopes on either side became higher and steeper, but were still apparently nearly bare of vegetation until we reached Chosica, about twenty-six miles from Lima, 2800 feet above the sea. At this place it was formerly the custom to halt for breakfast, but since the line has been re-opened, the only eatables to be found are the fruits, chiefly bananas and *granadillas*, which Indian women offer to the passengers.

Henceforward the line is fairly enclosed between the slopes on either hand, everywhere rough and steep, but, as is the nature of volcanic rocks, nowhere cut into precipices. The gradient becomes perceptibly steeper, being about one in thirty-three in the space between Chosica and San Bartolomé—about thirteen miles. Here the change of climate begins to be distinctly marked. It is evident that during a great part of the year the declivities are covered with vegetation, though now brown from drought, and they show the occasional action of running water in deep furrows and ravines. Here the engineers engaged on the railway first confronted the serious difficulties of the undertaking. Following the line from San Bartolomé to Chicla, the distance is only thirty-four miles, but the difference of level is 7317 feet, and the fifty-one miles between this and the summit-tunnel involve an ascent of 10,740 feet. The gradient is very uniform, never, I believe, exceeding one in twenty-six, the average being about one in twenty-eight. Some of the expedients adopted appear simple enough, though quite effectual for the intended purpose. Very steep uniform slopes have been ascended by zigzags, in which the train is alternately dragged by the locomotive in front, and then (the motion being reversed), shoved up the next incline with the engine in the rear. In one place I observed that we passed five times, always at a different level, above the same point in the valley below.

Among the more remarkable works on the line are the viaducts by which deep and broad ravines cut in the friable volcanic rocks have been spanned. The iron beams and girders that sustain these structures appear much slighter than I have seen used in Europe. In crossing one *barranca*, on what is said to be the loftiest viaduct in the world, I stood on the platform at the end of the car: there being no continuous roadway, the eye plunged directly down into the chasm below, over which we seemed to be travelling on a spider's web.

For a distance of about eight miles from San Bartolomé the railway keeps near to the bottom of the valley, between slopes whereon a distinct green hue is now visible, and some trickling rivulets are perceived in the channels of the ravines. On the opposite, or northern, slope are still distinctly seen the terraces by which in ancient days the industrious Indian population carried cultivation up the precipitous slopes to a height of more than fifteen hundred feet above the bed of the valley. For in this land, before the Spaniard destroyed its simple civilization and reduced the larger part to a wilderness, the pressure of population was felt as it now is in the southern valleys of the Alps. The fact that terrace cultivation commenced precisely in the part of the valley where we now find streamlets from the flanks of the high mountains above, which might be used for partial irrigation, tends to show that no considerable change of climate has occurred.

Before reaching the Surco station (forty-eight miles from Lima, and 6655 feet above the sea) the road finally abandons the Rimac, and commences the seemingly formidable ascent of the declivity above the left bank. For some distance a projecting buttress with a moderate slope enabled the engineers to accomplish the ascent by long winding curves; but before long we reached the first zigzags, which are frequently repeated during the remainder of the ascent. The tunnels are frequent, but fortunately for the most part short, as the rate of travelling is necessarily slow, and the artificial fuel used gives out black fumes of a stifling character. A further change of climate, welcome to the botanist, was now very obvious. Although the soil appears to be parched, it is clear that some slight rain must recur at moderate intervals. Vegetation, if not luxuriant, finds the needful conditions, and in the gardens of Surco tropical fruits, such as bananas, cherimolias, oranges, and granadillas, are cultivated with tolerable success. Of the indigenous plants in flower at this season the large majority were *Compositæ*, chiefly belonging to the sun-flower tribe (*Helianthoïdeæ*), a group characteristic of the New World.* It was tantalizing to see so many new forms of vegetation pass before one's eyes untouched. Most of them were indeed finally captured, but several yet remain as fleeting images in my memory, never fixed by closer observation.

About one p.m. we reached the chief village of the valley, San Juan de Matucana, fifty-five miles from Lima, and about 7800 feet above the sea. The train halted here for twenty minutes, and we discovered that very tolerable food is to be had at a little inn kept by an Italian. Hunger having been already stilled, the time was available for botanizing in the neighbourhood of the station, and, along with several cosmopolite weeds which we are used to call European, I found a good many types not before seen. Owing to the accident of having left my gloves in the carriage, I unwisely postponed to collect one plant not seen by me again during my stay in Peru. This was a small species of *Tupa*, a genus now united to *Lobelia*, with flowers of a lurid purple colour, which is said to have

---

* Of 138 genera of *Helianthoïdeæ* 107 are exclusively confined to the American continent, 18 more are common to America and distant regions of the earth, one only is limited to tropical Asia, and two to tropical Africa, the remainder being scattered among remote islands—the Sandwich group, the Galapagos, Madagascar, and St. Helena.

the singular effect of producing temporary blindness in those who handle the foliage, and I had been assured by Mr. Nation that he had verified the statement by experiment.* We were here in the intermediate zone, wherein many species of the subtropical region are mingled with those characteristic of the Andean flora. Hitherto the most prevalent families, after the *Compositæ*, had been the *Solanaceæ* and *Malvaceæ*. These have many representatives in the Andean flora, but henceforward were associated with an increasing proportion of types of many different orders.

As we continued the ascent in the afternoon our locomotive began to show itself unequal to the heavy work of the long-continued ascent, whether owing to defects in construction or, as seemed more probable, to the bad quality of the fuel supplied. Two stoppages occurred, required, as we learned, to clear out tubes. A considerable ascent was then achieved by a detour into a lateral valley above Matucana, returning to the Rimac at a much higher level, as is done on the Brenner line between Gossensass and Schelleberg.

Up to this time the scenery had fallen much below my anticipations. Owing to the nature of the rocks, there was an utter deficiency in that variety of colour and form that are essential elements in the beauty of mountain scenery. A still greater defect is the entire absence of forest. Along the course of the Rimac bushes or small trees, such as *Schinus molle*, two *Acacias*, *Salix Humboldtiana*, and others, are tolerably frequent; but on the rugged surface of the mountain

* See note to page 184.

slopes nothing met the eye more conspicuous than the columnar stem of a cactus, or dense rigid tufts of what I took to be a Bromeliaceous plant, most probably a species of *Puya*. The sun had set, and darkness was fast closing round us when the train came suddenly to a standstill, and the intelligent American guard informed us that a delay of at least twenty minutes was required to set the locomotive in working order.

The accident was in every way fortunate. We had just reached the Puente Infernillo, by far the most striking scene on the whole route, rendered doubly impressive when seen by the rapidly fading light. The railway had here returned to the Rimac, and is carried for a short distance along the right bank. In front the river rushes out of a narrow cleft, while on either hand the mountains rise to a prodigious height, with a steeper declivity than we had as yet anywhere seen. With a lively recollection of the Via Mala, the gorge of Pfeffers, and other scenes of a similar character, I could bring to mind none to rival this for stern sublimity. The impassable chasm that seemed to defy further advance, the roar of the river in the deeply cut channel below, the impending masses that towered up above us, leaving but a strip of sky in view, combined to form such a representation of the jaws of hell as would have satisfied the imagination of the Tuscan poet. To a botanist the scene awoke very different associations. Before it became quite dark I had captured several outposts of the Andean flora, not hitherto seen. The beautiful *Tropæolum tuberosum*, with masses of flowers smaller, but even more brilliant, than those of the common garden species, climbed over the bushes. A fleshy-leaved *Oxalis*, the first seen of a numerous group, came out of the crevices of the adjoining rocks, and *Alonsoa acutifolia*, which I had never seen but in an English greenhouse, was an additional prize.

Night had completely fallen as we resumed our journey, and although my curiosity was much excited in the attempt to follow the course of the line, I utterly failed to do so. Watching the stars as guides to our direction, where these were not cut off by the frequent tunnels, I could only infer that we were constantly winding round sharp curves, at times near the bottom of a deep ravine, with the roar of a torrent close at hand, and soon after working at a dizzy height along the verge of a precipice, with the muffled bass of a waterfall heard from out of the depths. Even after I had travelled the reverse way in broad daylight, I remained in some doubt as to the real structure of this part of the line. So far as I know, the first application of a spiral tunnel in railway construction was on the line across the Apennine between Bologna and Florence, but the spiral is there but a semicircle; you enter it facing north, and emerge in the opposite direction at a higher level. A similar device has been more freely resorted to in the construction of the St. Gothard line; but on this part of the Oroya line, completed before that of the St. Gothard was commenced, the spiral, if I mistake not, includes two complete circles, at the end of which the train stands nearly vertically above the point from which it started. It is by no means altogether a tunnel, as the form of a great projecting buttress has allowed the line to be carried in great part along a spiral line traced upon its flanks.

Nearly two hours after sunset we at length reached the terminus at Chicla, very uncertain as to the resources of that place in point of shelter and food. We had had the pleasure of meeting in the train Mr. H——, a distinguished German statesman, who had travelled with us in the *Islay* on his way from California to make the tour of South America. He was accompanied by Baron von Zoden, the German minister at Lima. As their object was merely to see the railway line, they intended to return on the following morning; but meanwhile we resolved to confront together any difficulties that might arise.

The architecture of Chicla is remarkably uniform, the only differences being in the size of the edifices. Stone, brick, tiles, slate, and mortar are alike unknown. Planks are nailed together around a framework, the requisite number of pieces of corrugated iron are nailed to some rafters on the top, and the house is complete. After stepping from the railway car and scrambling up a steep bank, we found ourselves before the chief building of the place, a so-called hotel, kept by a worthy German whose ill fortune had placed him on the borderland, where for some time the place was alternately occupied by small parties of Chilian or Peruvian troops. Besides some rooms on an upper floor occupied by the people of the house, the hotel consisted of two large rooms on the ground floor, where food and drink were supplied to all comers, with an adjoining kitchen. For such fastidious travellers as might require further sleeping accommodation than a cloak in which to roll themselves, and a floor on which to stretch their limbs, a long adjoining shed was provided. This was divided by thin partitions into four or five small chambers, each capable of holding two beds. Supper was before long provided; and when we afterwards learned the difficulties of our host's position, our surprise was excited more by the merits than by the defects of the entertainment.

We had been assured at Lima that, on going up to Chicla, we should be sure to suffer from the *soroche*, by which name the people of South America denote mountain-sickness, familiar to those who ascend from the coast to the plateau of the Andes. Knowing the height of Chicla to be no more than 12,220 feet above the sea, and never having experienced any of the usual symptoms at greater heights in Europe, I had treated the warning with derision so far as I was personally concerned, though not sure what effect the diminished pressure might have on my companion. I have described elsewhere* my experience at Chicla, which undoubtedly resulted from a mitigated form of mountain-sickness, the symptoms being felt only at night, and passing away by day and in exercise. They were confined to the first two nights, and after the third day, during which we ascended to a height of more than two thousand feet above Chicla, they completely disappeared.

With regard to mountain-sickness, the only matter for surprise, as it seems to me, is that it is not more

* In *Nature* for September 14, 1882.

frequently felt at lower elevations, and that the human economy is able so readily to adapt itself to the altered conditions when transferred to an atmosphere of say two-thirds of the ordinary density, where the diminished supply to the lungs is aggravated by the increased mechanical effort requisite to move the limbs, and raise the weight of the body in an attenuated medium. Observation shows that the effects actually produced at great heights vary much with different individuals, and that in healthy subjects the functions after a short time adapt themselves to the new conditions. It is obvious that this process must have a limit, which has probably been very nearly attained in some cases.

In spite of some statements lately published, I am inclined to believe that the utmost limit of height compatible with active exertion will be found to lie, according to individual constitution, between twenty and twenty-five thousand feet. As regards our experiences at Chicla, the difficulty is to account for the fact that the effects produced while the body is at rest should disappear during active exercise; and whatever the nature of the disturbance of the functions, this was not accompanied by any discernible derangement of the respiration or the circulation. It appeared to me that the seat of disturbance, such as it was, was limited to the nervous system.

On the evening of our arrival we met at the hotel the commandant of the Chilian detachment, and on presenting my letter from the commander-in-chief, he was profuse in offers of assistance. It was speedily arranged that we should start on the following morning, to ride as far as the tunnel at the summit of the pass to Oroya, where I promised myself an ample harvest among the plants of the higher region of the Andes. When morning broke, after a sleepless night with a splitting headache, I found or fancied myself unfit for a hard day's work; and, my companion being in much the same plight, we sent at an early hour to request that the excursion should be postponed till the following day. By the time, however, that we had dressed and breakfasted, the troubles of the night were all forgotten. A new vegetable world was outside awaiting us, and we were soon on the slopes above the station, where, in the person of my friend W——, I had the advantage of a kind and zealous assistant in the work of plant-collecting.

Deferring to a later page some remarks on the vegetation of the Cordillera, I need merely say that of this first delightful day the morning hours were devoted to the steep declivity of the mountain overhanging the left bank of the stream, while the afternoon was given to the less precipitous but more broken and irregular slopes on the opposite, or right, bank.

Having soon made the discovery that the supplies at Chicla were very limited, we had taken measures to procure a few creature comforts through the obliging conductor of the train, which left Chicla, in the morning, and was to return from Lima on the following evening. A far more serious deficiency was at the same time apparent. I had quite underrated the quantity of paper required to dry the harvest of specimens that I was sure to collect here, and no one but a botanist can measure the intensity of distress with which I viewed the prospect of losing precious specimens, and seeing shapes of beauty converted into repulsive masses of corruption, for want of the material necessary for their preservation. I addressed an urgent note to Mr. Nation, on whose sympathy as a brother naturalist I could safely count, telling him that unless I could find two reams of suitable drying-paper on my return, I should infallibly require accommodation in a lunatic asylum at Lima.

The scenery at Chicla is wild, but neither very beautiful nor very imposing. As in the lower valley of the Rimac, the slopes of the mountains are steep, but the summits are deficient in boldness and variety of form. Those lying on the watershed of the Cordillera, at a distance of fifteen or twenty miles, apparently range from seventeen to eighteen thousand feet in height, and on the first day of our visit showed but occasional streaks and patches of snow, while the sombre tints of the rocks exhibited little variety of hue even in the brightest sunshine.

Although the stream at Chicla is the main branch of the Rimac, its volume is here much reduced, not having yet received the numerous tributaries that fall into it between this place and Matucana. It is here no more than a brawling torrent, swelling rapidly after even a very moderate fall of rain, but prevented from ever dwindling very low by the snows, of which some patches at least remain at all seasons on the upper summits of the Cordillera. In a country without wood, and where the art of building in stone had made little progress, one of the most serious obstacles to any advance in civilization must have arisen from the difficulty of crossing the streams by which the upper ranges of the Andes are everywhere intersected.

The art of constructing suspension bridges must have originated in the subtropical zone of Eastern Peru, where the abundance of climbing plants with long, flexible, tough stems supplied the requisite materials. These, being light and easily transported, were everywhere used in the valleys of the Andes to sustain hanging bridges, of which the roadway was formed of rough basket-work. The only change that has resulted from the introduction of European arts is that of late years iron wire is used instead of flexible *lianes* to sustain the bridges ; but the roadway is still made of basket-work, which is rapidly worn by the feet of passing men and animals, and the natives have a disagreeable habit of stopping up the holes, not by mending the basket-work where this has begun to give way, but by laying a flat stone over the weak place. Being very slight and not nicely adjusted, these bridges swing to and fro under the feet of a passenger to an extent that is at first rather startling, but, as in everything else, habit soon makes one indifferent. Our first experience this afternoon was very easy, as the bridge connecting the station with the *pueblo*, or village of Chicla, was new and more solid than usual.

The little village, altogether composed of frail sheds, was occupied by the Chilian detachment of about two hundred men, posted here to guard the railway line. Four houses, larger than the rest, wherein the officers had established themselves, were adorned with conspicuous painted inscriptions worthy of the hotels of a great city. The *Fonda del Universo* informed the public that it contained "apartamentos para familias," and the rival establishments were no way inferior in the stateliness of their titles and the inducements offered. It must be recollected that Chicla is the first halting-place on the main, almost the only, line of communication between the coast and a magnificent region, as large as England, and teeming with natural resources—the *montaña* of Central Peru. Before the war the hostelries of Chicla were often crowded, and the accommodation doubtless appeared sumptuous to the wearied travellers who had been contending with the hardships of the journey from the interior, and the passage of the double range of the Andes.

I have already said that the supplies at our hotel were somewhat scanty. Inquiries for eggs were met by the reply that the Chilian soldiers had killed all the poultry, and milk was not to be thought of, because the cows had all been driven to a distance to save them from the Chilians. But these were only trifling inconveniences. The experience of our German landlord was full of graver matter. A foreigner in the interior of Peru during this abominable war is placed between the devil and the deep sea. Having no one to protect him, his property is at the mercy of lawless soldiery; he is an object of suspicion to both parties, and his life is in constant peril. Our host owed to a fortunate accident that he had not been shot by a Peruvian party under the suspicion of having given information to the enemy. He was certainly no lover of the invader; but, like every foreigner in Peru, he looked forward with undisguised dread to the day when the Chilians should depart.

If one had not recollected how very slowly and imperfectly the elementary rules of health have made way in Europe, it would have been hard to understand how men of education and intelligence, such as the great majority of the Chilian officers, should neglect the simplest precautions for preserving the health of themselves and their men. We had heard that the troops at Chicla had lost many men owing to a severe outbreak of typhoid fever, though the disease had recently almost disappeared. The cause was not far to seek. The ground all around the village was thickly strewn with the remains of the numerous baggage animals that had fallen from overwork, and the beasts that had been slaughtered by the soldiers. In South America the only sanitary officials are the carrion-eating birds. Near the coast the removal of offal is chiefly accomplished by the *gallinazo*, a large black vulture; in the Andes the condor takes charge of all carrion, and travels far in quest of it. It is likely that in the noisy neighbourhood of a detachment of soldiers the birds were shy of approach. If the remains had been dragged a short distance away from the village, they would have been quickly disposed of. As it was, the carcases were allowed to accumulate close to the sheds in which the men were lodged until they bred a pestilence. Things were mended, they said, at the time of our visit, yet, warned by vile emanations, I found the carcase of a horse lying close beside the *baraque* in which we slept; and it was only after energetic remonstrances that I succeeded in having it removed to some distance, where, doubtless, the condors made a savoury meal.

We were not curious to inquire too particularly what animal had supplied the material for our evening repast. It was enough that the skill of the Chinese boy who acted as cook had converted it into a very eatable dish. The work of the establishment seemed to be conducted altogether by two boys—the Chinese cook and a young German who acted as waiter. It was curious to notice that the intercourse between the two was carried on in English, or what passed as such. On many another occasion during my journey I observed the same thing. Throughout America, and I believe that the same is true in most countries out of Europe, English has become the *lingua franca*, the general medium of communication between people of different nationalities.

Having felt perfectly well all day, and inclined to believe that the discomforts of the previous night had arisen from some accidental cause, we had no hesitation in renewing the arrangement for an excursion to the *Tunnel en la cima*, and the Chilian commandant readily promised to send two horses, with a soldier who was to act as guide and escort, at seven o'clock on the following morning. Rather late, after some hours' work in laying out the plants collected during the day, I lay down to sleep, but in a short time awoke with a severe headache, accompanied by ineffectual nausea, the light supper being already digested. It was an undoubted case of mountain-sickness, which had to be borne through the sleepless dark hours until daylight summoned us to rise. As on the previous day, the operations of washing and dressing chased away the symptoms, and before seven o'clock we were ready to start. At half-past seven we began to lose patience, and despatched a messenger to ascertain the cause of delay. No answer coming, we resolved to go in quest of the promised steeds, and, shouldering the *impedimenta*, proceeded across the stream to the *pueblo*. We soon discovered that no order had been given the night before, and that the commandant had not yet made his appearance. The messenger had not ventured to awake him, and thought it safest to await events. Having discovered the high-sounding name of the "hotel" where he lodged, I lost no time in proceeding to the double-bedded room shared by our commander with a brother officer, and rousing them both from sleep. Profuse excuses in excellent Spanish, with a promise that not a moment should be lost, were but a poor salve for my growing impatience, though policy required some faint effort at politeness, which had to be maintained through what seemed intolerable and interminable delays, until we at last got under way at ten o'clock.

It was indeed aggravating to find an excursion, to accomplish which any naturalist would gladly traverse an ocean, maimed and curtailed by the indolence which is the curse of the American Spaniard. One circumstance, indeed, helped to moderate the keenness of my disappointment. Rather heavy rain had fallen throughout the night, and the mountains about the head of the valley, previously almost clear of snow, were now covered pretty deep down to the level of about fifteen thousand feet. I already judged that it would be difficult, starting so late, to reach the summit tunnel, if sufficient time were to be reserved for botanizing. With snow on the ground the vegetation would be concealed, and the chief interest of the expedition lost, so that I readily made up my mind that we should not attempt to reach the summit of the pass.

We had not gone far on the track when we came to a suspension bridge, over which our soldier-guide rode as a matter of course. Seeing the frail structure swing to and fro under the horse's feet, I confess that I felt much inclined to dismount and cross on foot; but in such cases one remembers that whatever men or animals are accustomed to do they are sure to do safely, and I rode on, admiring the judgment with which my horse avoided the weak places in the basket-work under his feet.

The track is well beaten, and in easy places broad and even; but here and there, where it climbs over some projecting buttress of rock, is rather rougher and steeper than I have ever seen elsewhere in mountain countries on a path intended for horsemen, excepting, perhaps, some choice spots in the Great Atlas. It was impossible to push on rapidly, for we overtook a succession of long trains of baggage-animals—mules, donkeys, and llamas—moving towards the interior at a rate of little over two miles an hour. As it was only in favourable places that it was possible to pass, our patience went through many severe trials.

At about thirteen thousand feet above the sea we passed two farmhouses, evidently constructed by European settlers, plain but neat in appearance, and the fields better kept than one could have expected in a spot so remote, each with a clump of well-grown trees of the Peruvian elder. Higher up the scenery was constantly wilder, desolate rather than grand, and with no trace of the presence of man until we reached Casapalta, a small group of poor sheds now occupied by an outpost of Chilian soldiers, nearly fourteen thousand feet above the sea.

We had now evidently reached the true Alpine region. At the head of the valley in front fresh snow lay on the flanks of the mountains where the dark rugged masses of volcanic rock were not too steep to allow it to rest, and the higher summits in the background were completely covered. The slopes near at hand were carpeted with dwarf plants thickly set, rising only a few inches from the surface. The only exception was an erect spiny bush, growing about eighteen inches high, with dark orange flowers, one of the characteristic Andean forms — *Chuquiraga spinosa*.

The guide seemed disposed to halt here, but we had not yet reached our goal, and we pushed on for about three miles, to a point about 14,400 feet in height, where it seemed judicious to call a halt. For some time the horses had begun to show symptoms of distress. The spirited animal which I rode panted heavily in ascending the gentle slope, and at last was forced to stop and gasp for breath every thirty or forty yards. Near at hand a slender stream had cut a channel through some rough rocks, and promised a harvest of moisture-loving Alpine plants ; and opposite to us, on the northern side of the valley, a wild glen opened up a vista of snow-covered summits, of which the more distant appeared to reach a height of about eighteen thousand feet.

It was now about one o'clock, and, our light early breakfast being long since forgotten, we hastily swallowed our provision of sandwiches formed of the contents of a sardine-box, which, flavoured with the pure cold water of the stream, seemed delicious. Although the sun which had shone upon us during the morning was now covered with clouds, and we were very lightly dressed, no sensation of cold was felt at this height, and I do not believe that the thermometer at any time during the day fell below 50°. Doubtless the feverish excitement of those unique two hours of botanizing in a new world left no space for sensitiveness to other influences. The mountain-sickness of the previous night was utterly forgotten, and no sensation of inconvenience was felt during the day.

Reserving some remarks on the botany of this excursion, there is yet to be mentioned here one plant of the upper region so singular that it must attract the notice of every traveller. As we ascended from Casapalta we noticed patches of white which from a distance looked like snow. Seen nearer at hand, they had the appearance of large, rounded, flattened cushions, some five or six feet in diameter, and a foot high, covered with dense masses of floss silk that glistened with a silvery lustre. The unwary stranger who should be tempted to use one of these for a seat would suffer from the experiment. The plant is of the cactus family, and the silky covering conceals a host of long, slender, needle-like spines, that penetrate the flesh, easily break, and are most difficult to extract. Unfortunately, the living specimen which I sent to Kew did not survive the journey.

At about three o'clock it was necessary to think of returning. Several precious plants had been passed on the way and remained to be collected, and it was only prudent to return to our quarters before night, which here falls so abruptly. Soon after we started along the descending track, a whirring sound overhead caused us to look up. Two magnificent condors swooped down from the upper region, and, wheeling round about forty feet above our heads, described a half circle, and, having satisfied their curiosity, soared again to a vast height, till they seemed mere black specks in the sky. Meanwhile my horse, fresh after the long halt, and apparently delighted at the prospect of returning to pleasanter quarters, broke into a gallop, and throughout the way it cost me some trouble to restrain his impatience.

As we drew near Chicla, there being yet half an hour of daylight, we dismounted and dismissed our guide with the horses, thus being able to secure several plants not seen elsewhere. One of these was a solitary plant of the common potato, growing in a wild place among dwarf bushes near the stream. I do not, however, attach any importance to the fact as evidence on the disputed question of the true home of a plant which in South America has been cultivated from remote antiquity. The valley of the Rimac has doubtless been a frequented highway since long before the Spanish conquest, and, as we know, the plant spreads easily in favourable conditions. As far as I know, all the evidence as to the plant being indigenous in Peru and Bolivia is open to suspicion, and the only part of the continent where it can be said to be certainly a native is Southern Chili and the sub-Alpine region of the Chilian Andes.

The excursion to the upper region apparently completed the work of acclimatization. We slept soundly, and no symptoms of *soroche* was afterwards experienced. When I sallied forth on the morning of the 23rd in quest of breakfast, which was made luxurious by a tin of Swiss milk received by the train from Lima, I found my friend W—— conversing in English with a Chilian officer. This gentleman, introduced as Captain B———, the son of English parents, was about proceeding in command of a small detachment to occupy some place beyond the Cordillera. The number of Englishmen in the Chilian service is not small, and there is no part of South America where the conditions of climate, the habits of life, and the character of the people seem to be so well suited to our countrymen.

One of the sights of Chicla was the daily despatch of trains of laden animals towards the interior. In the opposite direction the traffic was very limited, for since the war the working of the silver mines about Cerro de Pasco has been suspended, and little of the produce of the montaña now makes its way to the coast. But, war or no war, the wants of the inland population, living in a region which produces nothing but food and raw material, must in some measure be supplied. There was nothing very new in seeing goods packed on the backs of mules and donkeys, but the llamas and their ways were a continual source of interest. If the body be somewhat ungainly, the head with its large lustrous eyes may fairly be called beautiful. They vary extremely in colour. The prevailing hues are between light brown and buff, but we saw many quite white, and a few nearly black, with a good many mottled in large patches of white, and dark brown. The legs appear weak, and the animal can bear but a light burthen. On the mountain tracks, the load for a mule is three hundred pounds, that for a donkey two hundred pounds, while a llama can carry no more than a hundred pounds; and when any one attempts to increase the load, the animal lies down and moans piteously. He seems, indeed, not yet thoroughly resigned to domesticity, and there is a note of ineffectual complaint about his bearing and about all the sounds which he emits. One morning I was so much struck by what appeared to be the wailing of a child or a woman in distress, that I followed the sound until, behind a rock, I discovered a solitary llama that had somehow been separated from his companions. The advantage of the llama in the highlands of Peru, where fodder is scarce and must often be carried from a distance, is that he is able to shift for himself. Where the herbage is so coarse and so scanty that a donkey would starve, the llama picks up a living from the woody stems of the dwarf bushes that creep along the surface.

Supposing that most of the plants growing on the slopes around Chicla had been collected two days before, I expected to find it expedient to go to some distance from the village on the 23rd. But I had formed an inadequate idea of the richness of the Andean flora. Commencing with a ridge of rocks on the opposite side of the valley, only a few hundred yards from the ground before traversed, I found so many new and interesting forms of vegetation that at the end of three or four hours of steady work I had ascended only four or five hundred feet above the village, and I believe that ample occupation for a week's work to a collector might be found within one mile of the Chicla station.

As already arranged, we decided to return to Lima on the morning of the 24th of April. If other engagements had not made this necessary, the condition of my collections would have forced me to retreat. It was certain that without a speedy supply of drying-paper a large portion must be lost. As we were despatching an early breakfast, we were struck by the appearance of a tall, vigorous, resolute-looking man, booted up to the thighs, who had arrived during the previous night. He turned out to be a fellow-countryman, one of that adventurous class that have supplied the pioneers of civilization to so many regions of the earth. This gentleman had settled in the montaña of Eastern Peru, at a height of only about four thousand feet above the sea. His account of the country was altogether attractive, and it was only after entering into some details that one began to think that a man of a less cheerful and enterprising disposition might have given a less favourable report. The place which he has selected is only some twenty leagues distant from the river Ucayali, one of the great tributaries of the Maranon, which is destined hereafter to be the channel for direct water-communication between Eastern Peru and the Atlantic coast. At present the only obstacle to communication is the fact that the country near the river is occupied by a tribe of fierce and hostile Indians, who allow no passage through their country. The climate was described by our informant as quite delightful and salubrious, the soil as most fertile, suitable for almost all tropical produce, and many of the plants of temperate regions, and the supposed inconveniences as unimportant. Jaguars are, indeed, common, but the chief objection to them is that they make it difficult to keep poultry. Poisonous snakes exist, but the prejudice against them is unreasonably strong. No case of any one dying from snake-bite had occurred at our informant's location.

One drawback he did, indeed, freely admit. There was scarcely any limit to be set to the productive capabilities of the country, but, beyond what could serve for personal consumption, it was hard to say what could be done with the crops. He was then engaged in trying the possibility of transporting some of the more valuable produce of his farming to Lima. The journey had been one of extreme difficulty. In some of the valleys heavy rains had washed away tracks and carried away bridges, and he had been driven back to seek a passage by some other route. About one-half of his train of mules with their loads had been carried away by torrents, or otherwise lost; but our buoyant countryman, now virtually arrived at his journey's end, seemed to think the experiment a fairly successful one. He had received no news from

England since the beginning of the previous November, so that one or two newspapers five weeks old were eagerly accepted.

The return journey from Chicla to Lima was easy and agreeable, but offered little of special interest. I noticed a curious illustration of the effects of the sea-breeze on vegetation even at a distance of thirty or forty miles from the coast. As we descended, I observed that the acacias which abound in the middle zone of the valley were densely covered with masses of the white flowers of a climbing *Mikania*, quite masking the natural aspect of the shrub. I thought it strange that this appearance should not have struck me while on my way ascending the valley. On closer attention, I saw that the *Mikania* was entirely confined to the eastern side of the acacia, so that the same shrub, looked at from the western side, showed no trace either of the leaves or flowers of the visitor. On reaching the Lima station, I was kindly greeted by Mr. Nation, who at once relieved my most pressing anxiety by telling me that I should find two reams of filtering paper awaiting me at my hotel.

Having given in the twenty-second volume of the *Journal of the Linnæan Society* a list of the plants collected during my excursion in the Cordillera, it is needless to overload these pages with technical names, and I shall content myself with a few general remarks on the vegetation of this region, amidst which I passed a brief period of constantly renewed admiration and delight. In the first place, the general character of the flora of Chicla differed alogether from my anticipations, for the simple reason that the climate is completely different from what might, under ordinary conditions, be expected. I had seen reason to conjecture that, in ascending from the Pacific coast to the Cordillera, the rate of diminution of mean temperature would be less considerable than in most other parts of the world, but I was no way prepared to find it so slight as it really is. During the time of my visit, the mean temperature at Lima, 448 feet above the sea, was very nearly 70°, while the annual mean appears to be 66·6° Fahr.* The mean temperature at Chicla at the same season was estimated by me at 54°, with a maximum of 65·7°, and a minimum of 42°, and the first figure probably approximates to the annual mean. For a difference in height of 11,774 feet this would give an average fall of 1° Fahr. for 935 feet of elevation, or 1° Cent. for 512 metres; whereas, as

* The only accurate information that I have found respecting the climate of Lima is contained in a paper by Rouand y Paz Soldan, "Resumen de las Observaciones Meteorologicas hechas en Lima durante 1869," quoted in the French translation of Grisebach's "Vegetation du Globe." Reduced to English measures, they give the following results:—

Mean temperature of four years ...	66·6°	Fahr.
,, ,, January, 1869	74·3°	,,
,, ,, July, 1869	57·6°	,,
Rainfall in the year 1869	13·4	inches.
,, June, 1869	2·45	,,
,, July, 1869	2·72	,,
,, August, 1869	2·48	,,
,, September, 1869	2·33	,,
,, October, 1869	2·16	,,
,, remaining seven months	1·24	,,

There is reason to think that the temperature for July, 1869, given above was exceptionally low, and although the months during which fogs prevail are abnormally cool for a place within 13° of the equator, I believe that the thermometer rarely falls below 60° Fahr.

is well known,* the ordinary estimate found in physical treatises, resulting chiefly from the observations of Humboldt, would give for Equatorial America a fall of 1° Fahr. for about 328 English feet of increased altitude, or 1° Cent. for 180 metres. This rate of decrease would give a fall of 36·6° Fahr. in ascending from Lima to Chicla, whereas, as we have seen, the difference is probably little more than one-third, certainly less than one-half, of that amount. It is, therefore, with some astonishment that the stranger, arriving in this region of the Cordillera, finds himself amidst a vegetation characteristic of the Temperate zone,† and that many of the most conspicuous species are such as in mid-Europe require the protection of a greenhouse. Amongst the more attractive and characteristic of the Andean flora, I may mention five species of *Calceolaria*, *Alonsoa*, two fine *Loasaceæ* (one with large deep orange flowers and stiff hairs that penetrate the gloves, the other a climber with yellow flowers), several bushy *Solanaceæ*, and a beautiful clematis, which may hereafter adorn European gardens.

Along with many types of vegetation peculiar to the Andes, or more or less widely diffused throughout the Western continent, it was very interesting to a

---

* See Appendix A, On the Fall of Temperature in ascending to Heights above the Sea-level.

† It is a curious illustration of the utterly untrustworthy character of statements made by unscientific travellers to read the following passage in a book published by a recent traveller in South America, who visited Chicla in November, the beginning of summer. He declares that the fringe of green vegetation "dwindles and withers at a height of nine or ten thousand feet; . . . while on the upper grounds, where sometimes rain is plentiful, the air is too keen and cold for even the most dwarfish and stunted vegetation to thrive."

botanist from Europe to find so large a proportion of the indigenous plants belong to types which characterize the mountain vegetation of our continent. Of the genera in which the plants collected by me are to be classed, fully one-half belong to this category, and these genera include more than an equal proportion of species. I find, indeed, that fully sixty per cent. of the species in my collection belong to European genera, but that, with trifling exceptions, the species are distinct and confined to the Andean region. The reasonable conclusion is that the types which are thus common to distant regions must be of very great antiquity, and that the ancestors of the existing species must have spread widely at a very remote period of the world's history. Most of the plants in question belong to genera having very numerous species, of which it may be presumed that the parent forms possessed a strong tendency to variation.

The only tree seen at Chicla is a species of elder—*Sambucus Peruviana* of botanists—not widely differing from the common black elder of Europe.

Along with the numerous allies of the Old-World flora that characterize the indigenous vegetation, it was somewhat remarkable to find, in the upper valley of the Rimac, a number of cosmopolitan weeds, most of them common in Europe, which appear to have become thoroughly naturalized. Most of these, which are also found in the coast region of Peru, were undoubtedly introduced by the Spaniards; but there are a few, such as the common chickweed, whose wide diffusion throughout the world seems to me to be more probably due to transport by birds.

To the botanist, the most interesting features in the Andean flora are supplied by the great family of *Compositæ*. To this belong nearly one-fourth of all the plants collected by me, and nearly one-third of those found in the higher Alpine region ; and, as far as available materials allow me to judge, I believe these to be about the true proportions for the higher parts of the Andean chain. It is further remarkable that of the thirteen tribes into which the 780 genera and 10,000 species of this family have been divided, all but the two smallest tribes—*Calendulaceæ* and *Arctotideæ*—are represented in the Andes. To the European botanist, the most interesting group is that of the *Mutisiaceæ*, which is especially characteristic of the South American flora. Of 420 known species belonging to this tribe, fully 350 are exclusively American, the remainder being distributed through Australasia, and from South Africa to Southern Asia. They exhibit many unfamiliar forms very unlike what we are used to find elsewhere in the world. One of the first plants which I gathered was a tall, straggling climber with pinnate leaves ending in a tendril. I naturally thought of the vetch tribe, but I observed that the leaves were without stipules, and that the leaflets were not articulated to the midrib. Great, however, was my surprise when, on finding a flowering specimen, it revealed itself as a composite belonging to the genus *Mutisia*.

Next to the *Compositæ*, the grasses are of all the natural orders the most largely represented in the Andean flora, but with the difference that nearly all belong to genera common to the mountain regions of

Europe. The species are indeed different, but the general aspect does not strike the European botanist as presenting any marked features of novelty.

One further characteristic of the flora of Chicla is the great variety of species to be found within a small area. In this respect it seemed to me to rival the flora of Southern Spain and Asia Minor, which are known to be exceptionally rich in endemic forms. I am, of course, unable to judge whether in this part of the Andes the species are localized to nearly the same degree as in those parts of the Mediterranean region, and it is at least possible that the individual species which I saw crowded together at Chicla may have a relatively wide geographical range. The only social species, in some places covering large patches on the steep slopes, is a lupen growing in dense bushy masses.

Again guarding myself from the temptation to draw positive inferences from very slight opportunities for observation, I may add a few remarks on what I saw of the flora of the upper or Alpine zone of the Cordillera. This appears to be far more sharply defined at its lower limit than that which I shall designate as the temperate zone. In the latter, although the nights are at all seasons cool, actual frost is rarely experienced, and snow never lies on the ground. In the upper or Alpine zone, on the contrary, night frosts recur not unfrequently throughout the year, snow falls from time to time, more frequent in winter—from May to August—but does not lie long enough to provide a season of complete rest to the vegetative organs. To the influence of these conditions we may probably attribute the chief characteristics of the flora. With scarcely an exception, the species of this zone are stunted in growth, rising but a few inches from the surface, but have much developed prostrate or creeping woody stems, or underground rhizomes. Compared with the middle, or temperate, zone, the species generally belong to the same natural groups. Some of the families, however, which are characteristic of the middle zone, such as *Loasaceæ, Verbenaceæ*, and *Solanaceæ*, do not appear to reach the higher region.

Of forms characteristic of the Alpine region of mountains in the Old World I observed several; *e.g. Geranium, Astragalus, Valeriana, Draba*, a saxifrage, and a very small gentian.

To sum up my impressions as to the flora of the western slopes of the Cordillera, I should say that it appears to be naturally divided into three well-marked zones. The lower, or subtropical, extending to about eight thousand feet above the sea, characterized by deficient rainfall, moderate heat continued throughout the year, and a complete absence of cold, the thermometer rarely falling below 50°. The species here mainly belong to genera characteristic of the flora of tropical America, but, owing to the climatal conditions, are limited in number, and do not include groups requiring much moisture.

The middle, or temperate, zone, extending from about eight thousand to about thirteen thousand feet above the sea, possesses a very varied flora which includes many groups characteristic of the Andes, and entirely or mainly confined to that range, with representatives of numerous genera that are widely diffused through the temperate regions of the northern hemisphere, and a smaller number of representative species of groups belonging to the tropical American flora. The climate of this region is marked by the absence of all extremes of temperature. Cool nights, in which frosts are infrequent and of short duration, alternate with days wherein the shade temperature rarely surpasses 70°. The division between the temperate and subtropical zones is marked rather by the more frequent, though moderate, rainfall, which in the former recurs at intervals throughout the year, than by any marked change of temperature. Hence there may be distinguished a rather broad intermediate zone in which many of the characteristic forms of each meet and are intermingled; but this does not appear to be defined by any genera, or even by more than a few species peculiar to it, and does not deserve to be treated apart in a general survey of the flora.

The upper, or Alpine, zone of the Cordillera, extending from about thirteen thousand feet to the utmost limit of vegetation, is well defined by the circumstance that night frosts here recur throughout the year, and snow lies at least occasionally on the surface, while a somewhat greater amount of aqueous precipitation, in the form of rain or snow, combined with diminished evaporation, maintains a moderate degree of moisture in the soil. The proportion borne by some groups of the characteristic Andean flora as compared with the entire vegetable population is here larger than in the temperate zone, but other types better adapted to the climate of the latter zone are here nearly or altogether wanting. The forms common to the north temperate zone are present in about an equal proportion, while the representatives of the tropical flora are but very few.

With reference to the opinion expressed by writers of authority, and especially by Engler,* that the Andean flora is exceptionally rich in endemic genera and species, and to the explanation which would account for the facts, first, by the greater facility afforded for the extension of new varieties in dry climates, where the soil is not continuously covered by the existing vegetation; and, secondly, by the isolation of the summits, favouring the development of special local forms, I may venture on some sceptical remarks.

When we are struck by the large number of genera and species that are exclusively confined to the Andean flora, we are apt to forget the vast extent of the region which we are contemplating. Even if we exclude the mountains of Central America, and also those of Southern Chili, from Araucania to the Straits of Magellan, we have in the Andes a mountain region considerably more than three thousand miles in length, and from two hundred to over five hundred miles in breadth. This vast region is as yet far from being sufficiently explored to enable us to fix the geographical limits of its genera and species with any precision; but it appears to me that, while a very large number of genera are limited to the Andes as a whole region, the range of most of them within the limits of that region is very wide. I am further disposed to form a similar opinion as to the distribution

* "Versuch einer Entwicklungsgeschichte der Pflanzenwelt."

of the species if compared to what is found in some other mountain districts. If we were to find in South America anything like the variety of species limited to very small areas that is encountered in Southern Spain, Greece, Asia Minor, and Southern Persia, where on each mountain that we ascend we find several well-marked local species, differing from those in similar stations a few miles distant, the catalogue of the Andean flora would have to be extended to three or four times its actual length.

Fully agreeing, as I do, with Engler in his general conclusion that dry climates are more favourable than moist ones to the development of new varieties, which are the ancestors of future new species, I must remark that in the Andes, so far as we know, the species with very restricted area abound more in the upper zone, where the soil is relatively moist, than in the drier middle or lower zones. Nor does it appear that isolation of the summits can be with reason invoked as an explanation. The most marked feature in the range, and one that geologists have perhaps not taken enough to heart, is the extremely continuous character of the crest of the range, especially on the western side, as is evidenced by the fact that from Colombia to Southern Chili there are so very few passes below the limit at which snow frequently lies on the surface. For a rational explanation of the facts as to the distribution of mountain floras, we are forced to assume that the various agencies which are in daily operation—birds and land animals, winds, etc.—are competent to effect the transference of the great majority of species from one mountain to another not very far removed; and if that be true in districts where peaks are separated by arms of the sea or by intervals of low country having a very different climate, the process must be still easier in a chain so continuous as that of the Andes.

On the evening of the 24th I had the advantage of meeting the representatives of nearly all the European powers then present at Lima at the table of Don R. C——, a native gentleman of large fortune and influential position. The entertainment might properly be described as sumptuous, and, excepting in some royal palaces, could not easily be matched in Europe. One feature, indeed, was unique, and appealed to the susceptibility of a botanist. The vases heaped with choice specimens of tropical fruits could scarcely have been seen out of Peru. The occasion was not one on which political questions could with propriety be discussed, but I was struck by the complete agreement amongst men of various nationalities, whose duty it was to know the real state of things, as to the formidable prospect of anarchy and disorder that must ensue whenever the Chilian forces should be withdrawn from Lima and the adjoining provinces—a prospect, I need scarcely add, that has been since fully realized.

Soon after sunrise on the 25th Mr. Nation was good enough to call for me. We had agreed to make a short excursion along the bed of the Rimac, the best, if not the only, ground near the city where one can form some idea of the indigenous vegetation of the low country. As happens elsewhere, the river has carried down seeds or roots of many plants of the valley, which find a home on its broad gravelly bed, while the continual moisture has enabled many species of the plain, elsewhere dried up at this season, to maintain a vigorous growth. The little expedition was full of interest, and, with the aid of Mr. Nation's extensive local knowledge, I was able to make acquaintance with many forms of vegetation not hitherto seen. It was necessary to return early to the town, as my Chicla collections required many hours of diligent work until nightfall, when I had the pleasure of joining an agreeable party at the house of Mr. Graham, the British *chargé d'affaires*.

Among other scientific or social engagements, I called on the following day upon M. Lombardi, the author of a voluminous work on Peru, of which three large volumes have already appeared. M. Lombardi is a man of varied and extensive acquirements, especially in natural history, and in the course of frequent travels through the interior has accumulated a large mass of new materials of no slight value. Unfortunately, his work has been planned on a scale needlessly vast and costly; and now that the funds, at one time freely supplied by the Government, are no longer forthcoming, the prospect of its completion seems rather uncertain. The drawings and dissections of many species of plants from the higher regions of the Andes not hitherto figured, which M. Lombardi was good enough to show me, appeared to be very carefully executed, and their publication, in whatever form, would be welcomed by botanists.

I had accepted an invitation to visit on the 27th a *hacienda* belonging to Don R. C—— and his brothers at a place called Caudivilla, about twenty miles north of Lima. In company with an agreeable party of the officers of two Italian frigates then stationed at Callao, we started by the railway which runs parallel to the coast from Lima to Ancon and Chancay. At a station about three miles from the hacienda, we left the main line, and were conveyed to our destination on a private line of railway belonging to the estate. This is a tract of flat country about eight miles long by four in breadth, extending to the base of the outermost spurs of the Cordillera, and watered by a stream from the higher range in the background. It is almost exclusively devoted to sugar-cultivation, and in the large buildings which we inspected the whole process of extracting sugar and rum from the cane was proceeding on a large scale, and with the aid of the most complete machinery and apparatus. Although some fifteen hundred workmen are employed upon the works, it appeared as if human labour played but a small part in the processes wherein steam power was the chief agent. Trains of small trucks, laden with sugar-cane cut to the right length, were drawn up an incline, the contents of each tilted in turn into a huge vat, wherein it was speedily crushed. We followed the torrent of juice which constantly flowed from this reservoir through a succession of large chambers until it reached the final stage, in which, purified and condensed, it is at once converted into crystals of pure sugar when thrown off by the centrifugal action of a rapidly revolving axis, while the colourless pellucid product which is to furnish the rum of commerce was conveyed into vessels whose dimensions would put to shame the great tun of Heidelberg.

I confess to having felt less interest in the industrial results of this admirably conducted estate, than in what I was able to learn of the human beings employed and their relations to their employers; and I found here matter for agreeable surprise. The workmen are partly agricultural labourers engaged in the sugar-plantation and other outdoor work, partly those employed in and about the factory. Among them were representatives of various races, the Chinese being perhaps in a majority, but with a considerable proportion of negroes and half-caste natives of Peru. I was struck at first with a general air of well-being among all the working people, and I found this easily accounted for when I saw more of the arrangements made for their benefit.

Among other departments we were shown the hospital, small, but perfectly clean and airy, in which there were only three or four patients, and a school with a cheerful-looking young mistress surrounded by jolly-looking little children, who came forward unasked to display their acquirements in spelling. But what particularly pleased me was the large eating-house, or restaurant, where we found hundreds of workmen at their midday meal. They were not marshalled at long tables, but sitting in small groups round separate tables, every man chosing his own company, and calling for the dish which he preferred. Seeing these men, each with his napkin, enjoying his selected food, I could not help thinking that in the article of diet they are better off than a traveller in many parts of Europe, to say nothing of the population of the British Islands. I was assured that no profit whatever was made on this branch of the establishment. There was no pretence of philanthropy, but simply the intelligent view that as a mere matter of business it answered best that the working men should feel themselves to be well off. In point of fact, the mere threat to discharge a man from his employment is usually found to be sufficient to maintain order and industry.

There was little time available for botanizing here, and, the ground being all under cultivation, little of any interest to be found. On the way back I secured one of the beautiful reeds (*Gynerium*) which abound in tropical America. Herbarium specimens give little idea of a grass which, in moist situations, is from twenty to twenty-five feet in height, and whose flowering panicle is from four to five feet long.

On the following day, April 28, Mr. Nation again acted as my guide in a short walk about the outskirts of the city on the south and south-west sides. Nothing could be more uninviting than the appearance of the ground, which consists of volcanic sand, in most places completely bare of vegetation, but strewn with the refuse of the city, skeletons of cattle, and all sorts of *rejectamenta*, which make it the favourite resort of the black gallinazo (*Cathartes atratas*), the universal scavenger in this part of South America. The bird is deservedly protected by the population, which probably owes to its activity protection from pestilence. On the banks of some ditches and drains, and on some patches of waste land moistened by infiltration, we found several interesting plants. It was not evidence of the good character of the lower class in Lima to observe that on these occasions Mr. Nation carried a loaded revolver in his breast-pocket.

Amongst various items of information received from Mr. Nation, I was especially interested in the facts which he had observed in the neighbourhood of Lima regarding the disintegration of the exposed volcanic rocks. As he was kind enough to give me a written memorandum on the subject, along with specimens of the objects referred to, I think it better to give the substance in his own words.

"In one of the earlier editions of his ' Principles of Geology,' Sir Charles Lyell, on the authority of Mr. Cruikshank, speaks of the evidence afforded of a considerable rise of land in the neighbourhood of Lima by the appearance of the surface of hard green sandstone rocks hollowed out into precisely the forms which they assume between high and low water mark on the shores of the Pacific, while immediately below these water-worn lines are ancient beaches strewn with rounded blocks. One of these cliffs appears on the hill behind the Baños del Pingro, about seven hundred feet above the contiguous valley; another occurs at Amancaes, about two hundred feet above the sea;* and others at intermediate elevations.' Mr. Nation remarks that, having seen these appearances soon after his arrival at Lima, continued observation during more than twenty-five years has satisfied him not only that the hollows spoken of in the surface of the rocks

* The heights are certainly incorrect. The base of the hill of Amancaes is nearly seven hundred feet above sea-level, and Mr. Nation states that the two localities mentioned by Mr. Cruikshank are at about the same elevation.

are larger than they were, but that many new ones have been formed during the interval. He is satisfied that the appearances, which, he admits, exactly resemble those caused by the sea on shore rocks, are due to subaërial action. The chief agent, in his opinion, is a cryptogamic plant growing on the surface of the rock. During a great part of the year, when dense fogs prevail at this elevation, the plant is in active vegetation. In the alternations of relative dryness and dampness of the air the cells swell and mechanically remove scales from the surface, which are seen to accumulate rapidly in the course of a single season.

Having submitted a specimen of the cryptogam in question to the eminent lichenologist, Mr. Crombie, I am informed that the plant belongs to the group of lowly organized lichens, now distinguished as the *Ephebacei*, but formerly referred to the *Algæ*. In the absence of fructification, Mr. Crombie is unable to decide whether the specimen should be referred to *Sirosiphon* or *Spilonema*; but he is sceptical as to the possibility of any direct chemical action upon the rock arising from the growth of the lichen. Some indirect action may, in his opinion, be due to retention of moisture on surfaces covered by the lichen. This opinion is strengthened when it is remembered that the rock is not affected by carbonic acid, which might be derived from the air, or by vegetable acids which might be formed by the decomposition of the lichen. I am disposed to think that vicissitudes of temperature play a great part in the disintegration of rock surfaces, and such action must be increased by alternations of moisture and dryness which must occur where, during a great part of the year, the hills are covered with fog in the morning and exposed to the sun in the afternoon.

In connection with this subject I may remark that, in countries where the rainfall is very slight or altogether deficient, we are apt to be misled by the appearance of the surface, and to much overrate the real amount of disintegration. In the drier parts of the Mediterranean region, especially in Egypt, as well as in Peru and Chili, we constantly see rocky slopes covered with fine *débris* which represent the accumulated work of many centuries, remaining *in situ* because there is no agency at work to remove it, while in countries where the slopes are frequently exposed to the action of running water fresh surfaces are subjected to the action of the atmosphere, and the comminuted materials are carried to a distance to form alluvial flats, to fill up lakes, or ultimately to reach the sea-coast. A somewhat similar remark may be made with regard to rock surfaces habitually covered with snow and very rarely exposed to heavy rain. I have often observed in the Alps and Pyrenees that, when the snow disappears during the short summer of the higher regions, we generally find the surface covered with small fragments of the underlying rock, not removed by the slow percolation of water during the melting of the snow. The same phenomenon long ago attracted the attention of Darwin during his short excursion across the passes of the Chilian Andes.

I regretted much that my very short stay at Lima left me no time to visit the places where these curious appearances may be observed; but I trust that they may engage the attention of some future traveller more competent than myself to thoroughly investigate them.

The morning of the 29th of April, my last day in Peru, was fully employed in needful preparations. As is usual in South America, I was troubled by the dilatory habits of the natives. The passport, which was promised in the morning, and without which, as I was told, I should not be allowed to depart, was not forthcoming until late in the afternoon; and at length I went, after bidding farewell to my travelling companions and to some new friends, by the four-o'clock train to Callao, too late to have any time for visiting the surroundings of that curious place. The *Ayacucho* steamer of the Pacific Steam Navigation Company had already left her moorings, and lay in the outer harbour. Having hurried on board rather after the hour named for departure, I found that my haste was quite superfluous, as we were not under way till long after dark, about nine p.m.

I quitted Lima full of the interest and enjoyment of my brief visit, but full also of the sense of depression necessarily caused by the condition of a country whose future prospects are so dark. The ruinous war, and the occupation of the best part of Peru by a foreign army, are far from being the heaviest of her misfortunes. It may even be that they afford the best chance for her recovery. The immediate prospect is that of a feeble military despotism, tempered by anarchy. It seems possible that amongst the classes hitherto wealthy, and now reduced to comparative want, men of a type superior to the ordinary political adventurer may come forward ; some strong man, with resolute will and clear insight, may possibly arise, and re-establish order in the midst of a moral chaos ; but of such a deliverance there is as yet no promise. Conversing with men of very different opinions, I was unable to hear of any man whose name inspired confidence. Some such feeling had existed with regard to the President Pardo, but when he was assassinated no serious attempt was made to detect and punish his murderers. The only opinion which appeared to obtain general assent was that the worst of the adventurers who have been the curse of Peru was the late dictator Pierola.

One thing, at least, appears certain : if Peru is to be rescued from anarchy and corruption, it must be through the influence of a single will—by a virtual, if not a formal, autocracy. To believe that in such a condition of society as exists here progress can be accomplished by representative institutions seems to me as gross a superstition as the belief in the divine right of kings.

CHAPTER III.

Voyage from Callao to Valparaiso—Arica—Tocopilla—Scenery of the moon—Caldera—Aspect of North Chili—British Pacific squadron—Coquimbo—Arrival at Valparaiso—Climate and vegetation of Central Chili—Railway journey to Santiago—Aspect of the city—Grand position of Santiago—Dr. Philippi—Excursion to Cerro St. Cristobal—Don B. Vicuña Mackenna—Remarkable trees—Excursion to the baths of Cauquenes—The first rains—Captive condors—Return to Santiago—Glorious sunset.

THE voyage from Callao to Valparaiso was accomplished under conditions as favourable to the comfort and enjoyment of the passengers as that from Panama to Callao. The *Ayacucho* is a larger ship than the *Islay*, but built on a nearly similar plan, and except towards the end of the voyage, when we took on board a detachment of Chilian soldiers returning to Valparaiso, we had no inconvenience from overcrowding. I was very fully occupied in the endeavour to preserve and put away in good condition the rather large collections made during my stay in Peru. Notwithstanding the character of the climate, I found the usual difficulty felt at sea in getting my paper thoroughly dry, and for several days the work was unceasing. It had the effect of preventing my going ashore at two or three places which at the time appeared to me uninteresting, but which I afterwards regretted not to have visited.

By daylight on the morning of April 30 we were off Tambo de Mora, a small place near the mouth of the river Canete, which, at some seasons, is said to bring down a large volume of water from the Cordillera. After a very short stay we went on to Pisco, a more considerable place, but unattractive as seen from the sea, surrounded by sandy barren flats. It is, however, of some commercial importance, being connected by railway with Yea, the chief town of this part of Peru; and we remained in the roads about three hours, pursuing our voyage in the evening.

Our course on May 1 lay rather far from land, this being the only day during the voyage on which we did not touch at one or more ports. Under ordinary circumstances all the coast steamers call at Mollendo, the terminus of the railway leading to Arequipa, and thence to the highlands of southern Peru and the frontier of Bolivia. Arequipa being at this time occupied by a Peruvian force, and communication with the interior being therefore irregular and difficult, Mollendo was touched only on alternate voyages of the Pacific steamers.

I was impressed by the case of a Bolivian family on board which seemed to involve great hardship. An elderly father, with the manners and bearing of an educated gentleman, had taken a numerous family, chiefly young girls, with several servants, to Europe, to visit Spanish relations, and was now on his way to return to La Paz. The choice lay for him between the direct land journey from Arica, involving a ride of some two hundred miles through a difficult country, partly almost a desert, and partly through the defiles of the Cordillera, or returning by another steamer to Mollendo, and thence making his way between the hostile Chilian and Peruvian forces to the shores of his native lake of Titicaca. There was, in the latter case, the additional difficulty that Mollendo is about the worst port on the western coast of America. It is, in fact, an open roadstead, and, although there is little wind, the swell from the Pacific often breaks with a heavy surf upon the shore, and serious accidents are not infrequent. As all seamen are agreed, the terminus of the railway should have been fixed at Quilca, about the same distance from Arequipa as Mollendo, and, as usual in Peru, the selection of the latter is attributed to a corrupt bargain.

Early on May 2 we cast anchor opposite Arica. There is nothing deserving to be called a harbour; but a projecting headland on the south side of the little town protects the roadstead from the southerly breeze and the swell, which was here scarcely perceptible. On landing, I hastened along the shore on the north side, where a fringe of low bushes and some patches of rusty green gave promise to the botanist, and broke the monotony of the incessant grey which is the uniform tint of the Pacific coast from Payta to Coquimbo. As at very many other places on the coast, the maps indicate a stream from the Cordillera falling into the sea at Arica, but the traveller searches in vain for running water, or even for a dry channel to show where the stream ought to run. Nevertheless, Arica, unlike the places farther south, does actually possess fresh water in some abundance. The water from the Cordillera filters through the sandy belt of low country near the coast, and there are springs or wells sufficing not only for the local demand, but also for the wants of Iquique, a much more considerable place more than eighty miles distant. The little steamer whose office it is to carry the weekly supply of water to the Iquique people was taking her cargo on board at the moment, and one was at a loss to imagine what would happen if any mischance should befall the steamer or the engine. It is certain that under the intelligent rule of the Incas, many places now parched were made habitable by aqueducts carrying water from the mountains, and there are probably many other places where water might be procured by boring; but the porous character of the superficial soil makes this an uncertain resource, and the general uniformity of all the deposits gives little prospect of Artesian wells.

Near to the town are a few meagre attempts at cultivation in the shape of vegetable gardens, surrounded by ditches, into which it seems that a little water comes by infiltration. A few grasses and other herbaceous plants, mostly common tropical weeds, were to be found here. Elsewhere, the ground was, as usual on the coast, merely sand, with here and there clumps of bushes about six or seven feet in height, chiefly *Compositæ* of the characteristic South American genera, *Baccharis* and *Tessaria*. A bush of *Cæsalpinia Gilliesii*, with only a few of its beautiful flowers left, the ornament of hot-houses in Europe, struck one as a strange apparition on this arid coast.

The position of Arica, connected as it is by railway with Tacna, the centre of a rich mineral district, possessing the best anchorage on this part of the Pacific coast, and a constant supply of good water, must some day make it a place of importance. The headland which commands it is crowned by a fort, on which the Peruvians had planted a good many guns, and its seizure by the Chilians was one of the first energetic blows struck during the war.

For some reason, not apparent, the great waves which flow inland after each considerable earthquake shock have been more destructive at Arica than at any other spot upon the coast. Three times the place has been utterly swept away, and one memorial survives in the shape of the hull of a large ship, lying fully a mile inland, seen by us a few miles north of the town as we approached in the morning. On each occasion the little town has been rebuilt close to the shore. Experience has not taught the people to build on the rising land, only a few hundred yards distant. Each man believes that the new house will last his time—*Après moi le déluge*, with a vengeance!

At Arica the coast-line, which from the promontory of Ajulla, about 6° north latitude, has kept a direction between south-east and south-south-east for a distance of about twelve hundred English miles, bends nearly due south, and maintains the same direction for nearly double that distance. It is in the tract lying between Arica and Copiapò that the conditions which produce the so-called rainless zone of the Pacific

coast have had the maximum effect. In that space
of about six hundred miles (farther than from Liverpool to Oporto) there is no inhabited place—with the
possible exception of Pisagua—where drinkable water
is to be had. Nowhere in the world is there such an
extensive tract of coast so unfitted for the habitation
of man. But this same region is rich in products
that minister to human wants, and man has overcome
the obstacles that seemed to render them inaccessible.
Besides mines of copper, silver, and lead, the deposits
of alkaline nitrates, whose extent has not yet been
fathomed, richly reward the expenditure of labour and
capital. One after another industrial establishments
have arisen along the coast at places suitable for the
embarkation of produce, and some of these have
already attained the dimensions of small towns. The
Ayacucho called at no less than nine of these places,
and there are two or three others that are occasionally visited. At a few of them, as at Iquique, the
water-supply is partially or altogether conveyed by
sea, but most of them subsist by distillation from seawater.

As may well be supposed, there is little in these
places to interest a stranger, and a description of one
may serve for all. Some more or less extensive
works, with one or several tall chimneys, are the most
prominent feature. Near to each establishment are
three or four clean-looking houses for managers and
head agents, of whom the majority appear to be
English. Grouped in narrow sandy lanes near at
hand are the dwellings—mere sheds built of reeds—of
the working people. In some of the more consider-

able places an iron church, in debased sham-Gothic style, has been procured from the United States, and has been set up in a central position, with the outline of a *plaza* in front of it, and several drinking-shops clustered near.

The aspect of the coast is not less monotonous than that of the inhabited places. The sea-board is nearly a straight line running from north to south, and, except at Mejillones, I saw no projecting headland to break its uniformity. Nearly everywhere what appears to be a range of flat-topped hills from about eight to fifteen hundred feet in height, of uniform dull grey hue unbroken by a single patch of verdure, forms the background. In truth, these seeming hills are the western margin of the great plateau of the desert of Atacama, which at its edge slopes rather steeply towards the Pacific coast, sometimes leaving a level margin of one or two miles in width, sometimes approaching within a few hundred feet of the shore. I find it difficult to form a conception of the causes which have led to this singular uniformity in the western limit of the volcanic rocks of the plateau. Whether we suppose the mass to have been originally thrown out from craters or fissures in the range of the cordillera by subaërial or submarine eruptions, we should think it inevitable that the western front should show great irregularities corresponding to greater volume of the streams of eruptive matter in some parts.

Admitting—what may be held for a certainty—that, whatever may have been the original conditions, the whole region has since been submerged, and that marine action would have levelled surface inequalities, it is not easy to understand how the uniformity in the western front could have been brought about during the period of subsequent and comparatively recent elevation. If this had occurred along an axis of elevation near to the present coast-line, the effect must have been to produce a coast-range parallel to that of the Andes, with a watershed having an eastern as well as a western slope, and accompanying disturbance of the strata, such as we find on a great scale in western North America. Some indications of such action may be seen in Chili, south of Copiapò, and further to the south, but I am not aware of any fact to justify a similar supposition respecting this part of the coast of South America.

On the morning of May 3 we were anchored in front of Pisagua, which, being the port of Tarapacà, the chief centre of the nitrate deposits, is at present an active place. The houses are rather more scattered than usual, some of them being built on rising ground, apparently above the reach of earthquake waves. The range of apparent hills, fully fifteen hundred feet in height, rises steeply behind the little town, and the monotonous slope is broken by a long zigzag line marking the railway to Tarapacà. Some steep rocks rising from the sea to the south of the anchorage were in great part brilliantly white, recalling the appearance of quartz veins, or beds of crystalline limestone, dipping at a high angle. Thinking the existence of such rocks on this coast very improbable, I was anxious to inspect them; but when I was told that the time of our stay would merely allow of a short visit to the town, I did not care to land. The same appearances are common along the coast, and I soon afterwards ascertained that they are produced by the droppings of sea-birds—the same which, when accumulated in large masses, form the guano deposits of the detached rocks and islets of the coast.

In the afternoon we reached Iquique, which is, I believe, the largest of the unnatural homes of men on this coast. Some one who had gone ashore here returned, bringing copies of two newspapers, by which the public of Iquique are kept informed as to the affairs of the world. I had already seen with surprise, and had many further opportunities for observing, the extent to which the newspaper press in South America has absorbed whatever literary capacity exists in the country. Of information there is not indeed much to be gathered from these sheets; but of grand sentiments and appeals to the noblest emotions the supply seems inexhaustible. I regret to own that experience in other parts of the world had already made me somewhat distrustful of such appeals; but the result of my study of South American newspapers culminated in a severe fit of moral indigestion, and I do not yet receive in a proper spirit any appeal to the noblest sentiments of my nature.

I am far from supposing, however, that with those who read literature of this kind the debilitating effect attributed to it by some critics necessarily ensues. Some at least of the heroic virtues have survived. For a man to die for his country may not be the highest form of heroism, but in every age it has drawn forth the instinctive admiration of his fellows; and it is not at Iquique that one should think of making light of it. These waters, which, during the late war, witnessed the fight between the *Esmeralda* and the *Huascar*,* would, in another age of the world, have become as famous as those of Salamis.

On the morning of May 4 we called at Huanillos, a small place of recent growth, not marked on any map that I have seen. It lies within a few miles of the mouth of the Loa, which, as laid down on maps, appears to be a considerable stream, rising in the Cordillera and traversing in a circuitous course the Bolivian part of the Atacama desert. I naturally inquired why the mouth of such a river had not been selected as the site of a port. I was informed that, in spite of the maps, no water flows through the channel of the river, and that what can be obtained by digging is brackish and unfit for drinking. Whether this

* Two small Chilian wooden ships, the *Esmeralda*, of 850 tons, mounting eight guns, commanded by Arturo Prat, and the *Covadonga*, of 412 tons, with two guns, commanded by Condell, were engaged in the blockade of Iquique, when, on the 21st of May, 1879, they were attacked by the Peruvian ironclad *Independencia*, of 2004 tons, mounting 18 (chiefly heavy Armstrong) guns, commanded by J. G. Moore, and the monitor *Huascar*, of 1130 tons, mounting two 300-pounder Armstrong turret guns, besides two deck guns, under Miguel Grau, the most skilful and enterprising of the Peruvian commanders. The Chilian captains resolved on a desperate defence. After maintaining for two hours the fight against the *Huascar*, Arturo Prat resolved on the attempt to board his adversary. Bringing his ship alongside, he sprang on the deck of the *Huascar*; but the ships were separated at once, and two men only fell along with him, while the *Esmeralda* went to the bottom with her crew of 180 men, of whom several were picked up by the boats of the *Huascar*. The *Independencia*, following the little *Covadonga*, ran on the rocks in the shallows south of Iquique, and became a total wreck; while the *Covadonga*, though shattered by her enemy's guns, was able to reach Autofogasta. The heroism of the Chilian commanders saved their country, and at the critical moment changed the fortune of the war.

arises from the fact that the trials have been made too near the shore, within reach of the infiltration of sea-water, or that all the water traversing the region inland becomes impregnated with saltpetre, I am unable to decide ; but it seems probable that careful examination of the water, some of which undoubtedly finds its way underground from the Cordillera to the Pacific coast, might add considerably to the resources of the country. The cost of conveying water direct from the mountains to certain points in the interior, and thence to the coast, would possibly be repaid by the saving in fuel now used for the distillation of sea-water, to say nothing of the probability that some portions of the surface would become available for cultivation. The experience of the Isthmus of Suez, where a constantly increasing area near the course of the freshwater canal has become productive, should, I think, encourage the attempt.

About midday we reached Tocopilla, another place of recent creation, consisting of a large establishment with several chimneys and the usual group of sheds for the workmen. Steep rocky slopes rise close behind, and it seemed possible to see something of the conditions of life on this part of the coast without going beyond sight and hearing of the steamer. Being told that our stay was to be short, but that the steam-whistle would be sounded a first time exactly a quarter of an hour before our departure, I shouldered my tin vasculum and went ashore. Passing the houses, I at once steered for the rocky slopes behind. Here at last I found what I had often heard of, but in whose existence I had almost ceased to believe—a land absolutely without a trace of vegetable life. Among the dolomite peaks of South Tyrol I had often been told that such a peak was absolutely bare of vegetation, but had always found a fair number of plants in clefts and crevices. I had been told the same thing at Suez of the burnt-up eastward face of Djebel Attakah, where even on the exposed rocks I had been able to collect something; but here I searched utterly in vain. Not only was there no green thing; not even a speck of lichen could I detect, though I looked at the rocks through a lens. Even more than by the absence of life, I was impressed by the appearance of the surface, which showed no token that water had ever flowed over it. Every edge of rock was sharp, as if freshly broken, and on the steep slope no trace of a channel furrowed its face.* The aspect is absolutely that of the scenery of the moon—of a world without water and without an atmosphere. I saw no insect and no lizard, no living thing, with the

* In the preface to his "Florula Atacamensis," Dr. Philippi, who has explored this region more thoroughly than any other traveller, states that on the range of coast hills between the Pan de Azucar (lat. 26° 8' south) and Miguel Diaz (lat. 24° 36') the fogs, called in Peru *garua*, or *garruga*, deposit during a great part of the year some moisture which occasionally takes the form of fine rain, such as is familiarly known to occur on the hills near Lima. He remarks as singular the fact that the same phenomenon is not observed on the coast north or south of those limits. From more recent observations, it would appear that this is not strictly true as regards the higher coast hills near Coquimbo, but it seems to hold as regards the tract of coast to the northward, between the neighbourhood of Taltal and that of Iquique, a distance of about four degrees of latitude. It may be that the coast hills are lower here than further south, and that as the desert region inland rises very gradually, and has a higher temperature inland than near the coast, the formation of fog is prevented. Whatever be the cause, the absence of fog would go far to account for the utter sterility of this region.

K

strange exception that on the rocks nearest the houses there were several small birds, which appeared to be rather shy, and which I was not able to approach. I was afterwards told that these birds live on the grain which they are able to steal or to pick out from the manure in the stables, where a few horses and mules are kept for the needs of the place. Assuming this to be correct, the arrival of the birds at such a place remains a mystery.

A passenger who had spent some time at this singular place further told me that the horses, constantly fed on dry grain, and receiving but a scanty ration of distilled sea-water, usually become blind, but do not otherwise suffer in health. He added a story to the effect that some palings which had been painted green were found a few days after covered with marks of teeth, and with the paint almost completely removed. The mules, attracted by the colour, had sought the refreshment of green food, and had vainly gnawed away the painted surface.

However singular the aspect of nature in this place might be, it could not long detain a naturalist. A world without life is soon found to be monotonous; and after clambering about for some time, and satisfying myself that there was nothing to be found, I turned to the shore, where broken shells and other remains of marine animals presented at least some variety. Seaweeds appeared to be scarce, but some were to be seen in the little pools left among the rocks by the retreating tide.

Just as I was about to collect some objects which might have been of interest, the steam-whistle of the

Ayacucho summoned me to return to the ship. As I was by this time at some distance from the landing-place, I hurried back under a blazing sun, and reached the ship within less than twenty minutes, only to find that haste was quite superfluous, as we did not start until more than an hour later.

The sun had already set when we reached Cobija. This was, I believe, the first place inhabited on this part of the coast. Before the late war, Bolivia held the coast from the mouth of the Loa to the Tropic of Capricorn, a tract of about one hundred and sixty miles, rich in mineral wealth, the whole of which, along with the adjoining provinces of Peru, is now annexed to Chili. Cobija, which was a place of some importance, is now much reduced, and little business seems to be carried on there.

Early on the 5th of May we were before Antofagasta, now the most thriving place on this coast, if a place can be said to thrive which exists under such unnatural conditions. It is, however, slightly better off than its neighbours to the north. A gentleman who resided here for some time assured me that at intervals of five or six years a heavy fall of rain occurs here. At such times not only the coast region, but the Atacama desert lying between the Cordillera and the sea is speedily covered with fresh vegetation, which after a few months is dried up and disappears. At such times the guanacos descend from the mountains, and actually reach the coast.

We must, without my attention being called to the fact while in my cabin, have turned back to the northward after leaving Antofagasta, as in the dusk we were before Mejillones, which lies fully thirty miles north of the former place. It stands on a little bay, well sheltered from the south by a considerable rocky promontory, and, as I had been led to expect, the ground is here broken and irregular, offering more promise of safe retreat for the indigenous vegetation than anywhere else on this coast. I had looked forward with interest to an hour or two on this more promising ground, and it was a disappointment to be unable to profit by a comparatively long stay, for we remained at anchor after nightfall, embarking cargo and some passengers until midnight. For the third time within twenty-four hours we crossed the Tropic of Capricorn, and thenceforward remained in the south temperate zone. But in this region the term is in no way specially appropriate to the coast climate of Chili, for nothing can be more truly temperate than that of the so-called tropical zone which we were now leaving. During the voyage from Callao the thermometer properly shaded had but once (while anchored at Arica) reached 70° Fahr. It usually stood by night at 64° to 65°, and at about 68° by day, except occasionally when exposed to the cool southern breeze, when it fell rapidly on two occasions, marking only 62·2°.

My aneroid barometer by Casella, graduated only to 19 inches, and therefore useless during my visit to the Cordillera, did not appear to have suffered, as these instruments often do, by the reduced pressure. It did not vary during seven days by so much as one-twentieth of an inch from the constant pressure 29.9, and agreed closely with the ship's mercurial barometer. Perhaps, owing to the fact that my observations were not sufficiently frequent and recorded with sufficient accuracy, I failed to detect on this coast of America the daily oscillations of pressure, which in this latitude probably amount to about one-twentieth of an inch.

On the 6th of May we reached Taltal, a small place, the general aspect of which reminded me of Tocopilla, and my first impression on landing was that this was equally devoid of vegetable and animal life. But on reaching the rocky slope, which rises very near the landing-place, I at once perceived some indications of water having flowed over the surface, and in the course of the short half-hour which was allowed ashore I found three flowering plants, two of them in a condition to be determined, the third dried up and undistinguishable. In the evening we touched at Chañeral, a place rising into importance as being the port of a rich mining district. The southerly breeze had been rather stronger than usual during the afternoon, and some passengers complained of the motion of the ship. An addition of seventy tons of copper in the hold, which was shipped here by torchlight, appeared to have a remarkable effect in steadying the vessel.

We reached Caldera early on the 7th, and remained for five or six hours. This is the port of Copiapò, the chief town of Northern Chili—the only one, indeed, which could have grown up under natural conditions. A considerable stream, the Rio de Copiapò, which drains the western slope of the Cordillera, passes by the town. Caldera, the port, is not at the mouth of the river, but several miles further north, and water is doubtless conveyed there in some abundance, as, for the first time since leaving Arica, a few bushes in little enclosed gardens could be descried from the harbour, and I was afterwards shown two stately trees, the ornament of the place, which were nearly fifteen feet in height. I went inland about a mile and a half, visiting a slight eminence where the rock, evidently very recent, crops out at the surface, and one or two other promising spots. Most of the country was covered with sand, in places soft and deep, and anywhere else in the world I should have thought it wretchedly barren, but after my recent experience the meagre vegetation appeared almost luxuriant.

There is much interest attaching to the flora of this desert region of South-western America. The species which grow here are the more or less modified representatives of plants which at some former period existed under very different conditions of life. In some of them the amount of modification has been very slight, the species, it may be presumed, possessing a considerable power of adaptation. Thus one composite of the sun-flower family, which I found here, and also at Payta, is but a slight variety of *Encelia canescens*,* which I had seen growing luxuriantly in the gravelly bed of the Rimac near Lima, and along that river to a height of six thousand feet

---

* The four species of *Encelia* described in De Candolle's "Prodromus" appear to me to be but slightly modified forms of a single species. Since the publication of that work, several other and quite distinct species have been ranked under the same generic name.

above the sea. In this parched region the plant is stunted, and the leaves are hoary with minute white hairs, which may serve as a protection against evaporation. The same species, with other slight modifications, extends to all the drier portions of the western coast as far south as Central Chili. A dwarf shrub with yellow flowers like those of a jessamine, but with very different two-horned fruit, called *Skytanthus*, was an example of a much greater amount of change. Its only allies are two species in tropical Brazil, very different in appearance, though nearly similar in essential structure. We may safely conclude that a long period has elapsed since these forms diverged from a common stock, and that many intermediate links have perished during the interval.

Several of the ships composing the British Pacific squadron were lying at Caldera at this time, and after returning from my short excursion ashore, I went on board the *Triumph*, Captain Albert Markham, bearing the flag of Admiral Lyons, commander-in-chief. With regret I declined the admiral's hospitable invitation to accompany the squadron to Valparaiso, but I was unable to refuse Captain Markham's kind suggestion that, as his ship was under orders to return to England on the arrival of the *Swiftsure*, then expected, I should become his guest on the passage from Valparaiso to Montevideo. The *Triumph* having been detained in Chilian waters many weeks longer than was then expected, I was afterwards forced to forego the agreeable prospect of a voyage in company with an officer whose varied accomplishments and extensive observation of nature under the most varied conditions make his society equally agreeable and instructive.

Leaving Caldera soon after midday, the *Ayacucho* reached Coquimbo early on the following morning. With only the exception of Talcahuano, this is the best port in Chili, being sheltered from all troublesome winds, and affording good anchorage for large ships. The town of La Serena, the chief place in this part of Chili, stands on moderately high ground about two miles from the sea, and may be reached in about twenty minutes from the port by frequent trains which travel to and fro. We were warned that our stay was to be very short, and that those who went to the city could not remain there for more than half an hour. I had no difficulty in deciding to forego the attractions of the city, whatever they might be, for a far more tempting alternative offered itself. The range of low but rather steep slopes that rises immediately behind the chief line of street was actually dotted over with bushes, veritable bushes, and the unusual greenish-grey tint of the soil announced that it was at least partially covered with vegetation. In the spring, as I was assured, the hue is quite a bright green. To a man who for the preceding week had seen nothing on land but naked rocks or barren sand, the somewhat parched and meagre vegetation of Coquimbo appeared irresistibly attractive. I could not expect to add anything of value to what is already known, through the writings of Darwin and other travellers, respecting the evidences of elevation of the coast afforded by the raised terraces containing recent shells, whose seaward face forms the seeming hills behind the town, and I felt free to give every available moment to collecting the singular plants of this region.

One of the minor satisfactions of a naturalist in South America arises from the fact that the inhabitants are so thoroughly used to seeing strangers of every nationality, and in the most varied attire, that his appearance excites no surprise and provokes no uncivil attentions. Going about almost always alone, with a large tin box slung across my back, I never found myself even stared at, which, in most parts of Europe, is the least inconvenience that befalls a solitary botanist. The amount of attention varies, indeed, in different countries. In Sicily and in Syria one is an object of general curiosity, and one's every movement, as that of a strange animal, watched by a silent crowd; but it is only in Spain that the inoffensive stranger is subject to personal molestation, and the little boys pelt him with stones without rebuke or interference from their seniors, who nevertheless boast of their national courtesy.* Fortunately it nowhere occurs to the most ill-disposed populations that a shabbily dressed man, engaged in grubbing up plants by the roots, can be worth robbing. Usually regarded as the assistant to some pharmacist, the botanist is, I think, less subject to molestation than

* While botanizing in the Tajo de Ronda, the singular cleft which cuts through the rocky hill on which the town is built, I was once for some time in positive danger. The boys, having espied me, assembled on the bridge that crosses the cleft, some three hundred feet above my head, and commenced a regular fire of stones, that drove me to take shelter under an overhanging rock until, being tired of the sport, they turned their attentions elsewhere.

the follower of any other pursuit; his only difficulty being that, if ignorant of the healing art, he cuts a poor figure when applied to for medical advice.

Quite unnoticed, I made my way through the long street of Coquimbo, and, at the first favourable opportunity, turned up a lane leading to the slopes above the town. The first plant that I saw, close to the houses, was a huge specimen of the common European *Marrubium vulgare*, grown to the dimensions of a much-branched bush four or five feet high. It is common in temperate South America, reaching a much greater size than in Europe. The season was, of course, very unfavourable, the condition of the vegetation being very much what may be seen at the corresponding season—late autumn—in Southern Spain, before the first winter rain has awakened the dormant vegetation of the smaller bulbous-rooted plants. Nevertheless, I found several very curious and rare plants still in flower, some of them known only from this vicinity, and among them a dwarf cactus, only three or four inches in height, with comparatively large crimson flowers just beginning to expand.

At length, on the morning of May 9, the voyage came to an end as we slowly steamed into the harbour of Valparaiso, which, with the large amount of shipping and the conspicuous floating docks, gives an impression of even greater importance than it actually possesses. The modern town, built in European fashion, with houses of two and even three floors above the ground, on the curved margin of the bay partly reclaimed from the sea, and the older town, chiefly perched on the edge of the plateau some two hundred feet above the main street, and divided by the deep ravines (*quebradas*) that converge towards the bay, have been described by many travellers; but I do not remember to have seen any sufficient warning as to the frightful peril to which the majority of the population is constantly exposed. Over and over again earthquakes have destroyed towns in western South America. Houses built of slight materials, with a ground floor only, or at most with a single floor above it, may fall without entailing much loss of life; but it is frightful to contemplate the amount of destruction of life and property that must ensue if a violent shock should ever visit Valparaiso. And the peril is twofold; the great wave which is the usual sequel of a violent earthquake, would inevitably destroy whatever might survive the first shock in the crowded streets of the lower town.

After overcoming the preliminary difficulties of landing and passing my luggage through the custom-house, I proceeded to the Hotel Colon, in the main street, kept by a French proprietor to whose lively conversation I owed much information and amusement during my short stay. Some three hours were occupied by a few visits, a stroll through the chief streets, and the despatch of a telegram to Buenos Ayres. Not choosing to incur the heavy expense of a telegram from Valparaiso to England, I had availed myself of the courtesy of the officials of the Royal Mail Steamboat Company to arrange that a telegram from Valparaiso to Buenos Ayres should be forwarded by post from the latter place, thus saving fully three weeks' time. In the afternoon I climbed up one of the steep tracks leading to the upper part of the town, where the population mainly consists of the poorest class. The houses were small and frail looking, but fairly clean, and I nowhere saw indications of abject want, such as may too often be witnessed in the outskirts of a large European city.

Valparaiso has all the air of a busy place, with some features to which we are not used in Europe. Along the line of the narrow main street tramcars are constantly passing to and fro. The names over the shops, many of which are large and handsome, are mainly foreign, German being, perhaps, in a majority; but the important mercantile houses are chiefly English, and, except among the poorer class, the English language appears to be predominant. All people engaged in business acquire it when young, and very many of Spanish descent speak it with fluency and correctness. The Hotel Colon stands between the main street and a broad quay, part of the space reclaimed from the margin of the bay, and my windows overlooked the busy scene, thronged from daylight till evening with a crowd of men and vehicles. It was somewhat startling to see frequent railway trains run through the crowd, with no other precaution than the swinging of a large bell on the locomotive to warn people to get out of the way.

I started soon after daylight on the 10th for a botanical excursion over the hills behind the town, and, as I had rather exaggerated expectations of the harvest to be collected, I had engaged a boy to carry a portfolio wherein to stow away what I could not conveniently carry for myself. Though I moved slowly, as naturalists generally do, my companion soon grew tired, or pretended fatigue, and after an hour or so I sent him back to the hotel with the portfolio well filled.

The flora of Central Chili is denominated by Grisebach that of the transition zone of western South America; but, except in the sense that it occupies a territory intermediate between the desert region to the north and that of the antarctic forests to the south, the term is not very appropriate. On the opposite side of the continent, the flora of Uruguay, Entrerios, and the adjoining provinces, may be truly said to offer a transition between that of South subtropical Brazil and that of the pampas region, most of the genera belonging to one or other of those regions, the one element gradually diminishing in importance as the other assumes a predominance. In this respect the Chilian flora presents a remarkable contrast, being distinguished by the large number of vegetable types peculiar to it, and having but slight affinities either with those of tropical or antarctic America.

Of 198 genera peculiar to temperate South America, the large majority belong exclusively to Central Chili, and these include several tribes whose affinity to the forms of other regions is only remote. Two of these tribes—the *Vivianeæ* and *Francoaceæ*—have even been regarded by many botanists as distinct natural orders; and many of the most common and conspicuous species will strike a botanist familiar with the vegetation of other regions of the earth as very distinct from all that he has known elsewhere.

With only a few exceptions, these endemic types appear to have originated in the Andean range, whence some modified forms have descended to the lower country; several of these, as was inevitable, have been found on the eastern flanks of the great range, and it is probable that further exploration will add to the number; but it is remarkable that as yet so large a proportion should be confined to Chilian territory.

Grisebach has fixed the limits of that which he has called the transition zone at the Tropic of Capricorn to the north, and the thirty-fourth parallel of latitude to the south; but these in no way correspond to the natural boundaries. As I have already pointed out, the flora of the desert zone, extending from about the twentieth nearly to the thirtieth parallel of south latitude, shows a general uniformity in its meagre constituents. It is about the latitude of Coquimbo, or only a little north of it, that the characteristic types of the Chilian flora begin to present themselves, and these extend southward at least as far as latitude 36° south, and even somewhat farther, if I may judge from the imperfect indications of locality too often afforded with herbarium specimens.*

---

* One of the difficulties felt by all students of geographical distribution arises from the imperfect or careless indications given both in books and in herbaria, and this is more felt in regard to South America than as to any other part of the world. A very large proportion of the earlier collections bear simply the label "Brazil," forgetting that the area is as great as that of Europe. In other cases local names of places, not to be found on maps or in gazetteers, embarrass the student and weary his patience. It is mainly from Darwin that naturalists have learned that geographical distribution is the chief key to the past history of the earth.

In discussing the causes that have operated on the development of the Chilian flora, the same eminent writer has been misled by incomplete and erroneous information as to the climate of the region in question, and more especially as to the distribution of rainfall, which is no doubt the most important factor. It is true that the peculiar character of the Chilian climate makes it very difficult to express by averages the facts that mainly influence organic life. Between the northern desert region, where rain in a measurable quantity is an exceptional phenomenon, and the southern forest region, extending from the Straits of Magellan to the province of Valdivia, Central Chili has in ordinary years a long, dry, rainless summer, followed by rather scanty rainfall at intervals from the late autumn to spring. About once in four or five years an exceptional season recurs, when rain falls in summer as well as winter, and in which the total fall may be double the usual amount, and at longer intervals, usually after a severe earthquake, storms causing formidable inundations occur, when in a few days the rainfall may exceed the ordinary amount for an entire year. When several such storms are repeated in the same year, we may have a total rainfall of three or four times the ordinary average.*

* The last season of excessive rainfall was that of 1877. I have seen no complete returns, but it appears that the rain of that year commenced in Central Chili in February, a very rare phenomenon ; that more than six inches of rain fell in April, of which, at Santiago, four inches fell in twenty-four hours. More heavy rain fell in May, and finally in July a succession of storms flooded large districts, destroying property and life, the fall for the month being more than fourteen inches at Valparaiso. Much interesting information respecting the climate of Chili will be found in a work by Don B. Vicuña Mackenna, " Ensayo

Such seasons appear to recur six or seven times in each century, and it is clear that, according as the meteorologist happens to include or exclude such a season in his data, the figures expressing the average must vary very largely. Inasmuch as plant life is regulated by the ordinary conditions of temperature and moisture, we are less liable to error in taking the results which exclude exceptional seasons.

In discussing, therefore, the conditions of vegetation in Central Chili, it seems safe to conclude that the averages given in the following table, extracted from the careful work of Julius Hann, "Lehrbuch der Klimatologie," are above rather than below the ordinary limit. I find, indeed, that while the average rainfall at Santiago during the twelve years from 1849 to 1860 was 419 millimetres, or nearly $16\frac{1}{2}$ English inches, the average for the six years from 1866 to 1871 was 299 millimetres, or less than 12 inches. It is evident that the indigenous vegetation must be adapted to thrive upon the smaller amount of moisture expressed by the latter figures.

The following table, compiled from Hann's work, gives the most reliable results now available, and shows the mean temperature of the year, of the hottest and coldest months, the extremes of annual temperature, and the rainfall for the chief places in Chili, with a few blanks where information is not available. The maxima and minima do not express the absolute extremes attained during the entire period for which observations are available, but the means of

Historico sobre el Clima de Chile" (Valparaiso: 1877), from which I have borrowed the above-mentioned particulars.

the annual maxima and minima. The temperatures are given in degrees of Fahrenheit's thermometer.

Places.	South latitude.	Mean temperature of the year.	Mean temperature of January.	Mean temperature of July.	Maximum temperature.	Minimum temperature.	Rainfall in inches.
La Serena (Coquimbo)	29° 56′	59·2°	65·1°	53·1°			1·6
Valparaiso	33° 1′	57·6°	63·1°	52·5°	77·9	45·0°	13·5
Santiago (1740 feet above sea-level)	33° 27′	55·6°	66·2°	45·0°	87·6°	30·4°	14·5
Talca (334 feet above the sea)	35° 36′	56·5°	70·2°	45·0°			19·7
Valdivia	39° 49′	52·9°	61·5°	45·0°	84·0°	29·5°	115·0
Ancud, Island of Chiloe (164 feet above the sea)	41° 46′	50·7°	56·5°	45·9°			134·0
Punta Arenas * (Straits of Magellan)	53° 10′	43·2°	51·3°	34·9°	76·3°	28·4°	22·5

This table brings out very clearly the influence of the cold southern currents of the ocean and air in reducing the summer heat of the western side of South America; for, while the winter temperatures are not very different from those of places similarly situated on the west side of Europe and North Africa, those of summer are lower by 8° or 10° Fahr., and the mean of the year is lower by 6° or 7° than that of places in the same latitude on the east side of South America. It is also apparent that much of what has been stated in works of authority as to the climate of this region is altogether incorrect. In his great work on the "Vegetation of the Earth," Grisebach gives the mean temperature of Santiago as 67·5°, or nearly 12° higher than the mean result of ten years' observation, and the rainfall as over 40 inches, or

---

* I believe that in the column for rainfall at Punta Arenas, snow has not been taken into account.

nearly three times the average—more, indeed, than three times the average—of ordinary seasons.

Arriving in Chili about the end of the long dry season, I had but very moderate expectations as to the prospect of seeing much of its peculiar vegetation, and I was agreeably surprised to find that there yet remained a good deal to interest me, especially among the characteristic evergreen shrubs, having much of the general aspect of those of the Mediterranean region, though widely different in structure from the Old-World forms. One or two slight showers had fallen shortly before my arrival, and as a result the ground was in many places studded with the golden flowers of the little *Oxalis lobata*. This appears to have a true bulb, formed from the overlapping bases of the outer leaves, in the centre of which the undeveloped stem produces one or more flowers, which appear before the new leaves. The surface of the dry baked soil was extremely hard, costing some labour to break it with a pick in order to collect specimens, and it is not easy to understand the process by which a young flower-bud is enabled to force its way to the upper surface. The open country on the hills near Valparaiso is bare, trees being very scarce, and for the most part reduced to the stature of shrubs with strong trunks; but in the ravines, or *quebradas*, that descend towards the coast some of these rise to a height of twenty or twenty-five feet.

One of the objects of my walk over the hills was to obtain a good view of the Andes, and especially of the peak of Aconcagua, the highest summit of the New World. I had had a glimpse of the peak from the sea on the previous morning, but light clouds hung about the entire range during this day, and I was unable to identify with certainty any of the summits. The distance in an air line is about one hundred English miles, and I was struck by the clearness of the air in this region as compared with what I had seen from the coast of Ecuador or Peru. Every point that stood out from the clouds was seen sharply defined, as one is accustomed to observe in favourable weather in the Mediterranean region.

Returning to the town, I took my way along one of numerous deep ravines that have been cut into the seaward surface of the plateau. Though they are witnesses to the energetic action of water, they are often completely dry at this season; yet they exercise a marked influence on the vegetation. The shrubs rise nearly to the dimensions of trees, and several species find a home that do not thrive in the open country. I was specially interested in, for the first time, finding in flower the Winter's bark (*Drimys Winteri*), a shrub which displays an extraordinary capacity of adaptation to varying physical conditions, as it extends along the west side of America from Mexico to the Straits of Magellan, and also to the highlands of Guiana and Brazil, accommodating itself as well to the perpetual spring of the equatorial mountain zone as to the long winters and short, almost sunless, summers of Fuegia. The only necessary condition seems to be a moderate amount of moisture; but even as to this there is wonderful contrast between the long rainless summer and slight winter rainfall of Valparaiso, and the tropical rains of Brazil on the one hand, and the continual moisture of Valdivia and Western Patagonia on the other. This is one of the examples which goes to show how much caution should be used in drawing inferences as to the climate of former epochs from deposits of fossil vegetable remains. This instance is doubtless exceptional, but there is some reason to think that what may be called physiological varieties—races of plants which, with little or no morphological change, have become adapted to conditions of life very different from those under which the ancestral form was developed—are far less uncommon than has been generally supposed.

It is to me rather surprising that a shrub so ornamental as the Winter's bark should not be more extensively introduced on our western coasts. It appears not to resist severe frosts, but in the west of Ireland and the south-west of England it should be a welcome addition to the resources of the landscape gardener. Although voyagers have spoken highly of its virtues as a stimulant and antiscorbutic, it does not appear to have held its ground in European pharmacopœias, and I believe that the active principle, chiefly residing in the bark, has never been chemically determined.

On May 11 I proceeded to Santiago. Mr. Drummond Hay,[*] the popular consul-general, who at this time was also acting as the British *chargé d'affaires* at the legation at Santiago, was so fully occupied at the consular court that I was able to enjoy little of his society; but he was kind enough to telegraph to the

[*] The recent untimely death of this valuable official is deplored by all classes in Chili.

Hotel Oddo at Santiago to secure for me accommodation. With the usual difficulty of effecting an early start, which appears to prevail everywhere in South America, I reached the railway station in time for the 7.45 a.m. train. For some distance the railroad runs near the sea, passing the station of Viña del Mar, where many of the Valparaiso merchants have pretty villas. I was more attracted by the appearance of the country about the following station of Salto, where rough, rocky ground, with clumps of small trees and the channels of one or more streams, promised well for a spring visit. But I was at every turn reminded that I had fallen on the most unfavourable season. After the long six or seven months' drought the face of the country was everywhere parched, and the only matter for surprise was that there should yet remain some vestiges of its summer garb of vegetation.

The direct distance from Valparaiso to Santiago is only about fifty-five miles, but the line chosen for the railway must be fully double that length. The country lying directly between the sea-coast and the capital is broken up by irregular masses, partly granitic and partly formed of greenstone and other hard igneous rocks. These in Europe would be regarded as considerable mountains, as the summits range from six thousand to over seven thousand feet in height, but they nowhere exhibit the bold and picturesque forms that characterize the granite formation in Brazil. On either side of this highland tract two considerable streams carry the drainage of the Cordillera to the ocean. The northern stream, the Rio Aconcagua, bears the same name as the famous mountain from whose snows it draws a constant supply even in the dry season. Some sixty miles further south, the Maipo, draining a larger portion of the Andean range, flows to the coast by the town of Melipilla. The valley of the Maipo offers a much easier, though a circuitous, railway route to Santiago than that chosen by the Chilian engineers, which for a considerable distance keeps to the valley of the Aconcagua. The stream is reached near to Quillota, a place which has given its name to this part of the valley.

Travelling at this season, I was not much struck by the boasted luxuriance of the vegetation of the vale of Quillota; but I could easily understand that the eye of the stranger, accustomed to the arid regions of Peru and Northern Chili, must welcome the comparative freshness of the landscape, in which orchards of orange and peach trees alternate with squares of arable land. Of the few plants that I could make out from the railway car what most attracted my attention was the frequent recurrence of oval masses of dark leaves, much in the form of a giant hedgehog three or four feet in length and half that height, remarkably uniform in size and appearance. The interest was not diminished when I was able, at a wayside station, to ascertain that the plant was a bramble, on which I failed to find flower or fruit, but which from the leaves can be nothing else than a variety of the common bramble, or blackberry, introduced from Europe.

At the station of Llaillai (pronounced *Yaiyai*) we met the train from Santiago, and were allowed a quarter of an hour for breakfast. The arrangements were rather rough, but the food excellent—much superior, indeed, to what one commonly finds at an English refreshment-room. This is a junction station, and a train was in readiness to take passengers from Santiago or Valparaiso by a branch line up the valley of the Aconcagua to San Felipe and Santa Rosa de los Andes. The Santiago train here leaves that valley, and, turning abruptly to the south, commences a long and rather steep ascent of the ridge that divides the basin of the Maipo from that of the Aconcagua. To our right rose the Cerro del Roble, about 7250 feet in height, one of the highest of the coast range.*

Here I first encountered the characteristic aspect of the hilly region of Central Chili. A tall columnar cactus (*Cereus Quisco*) is the most conspicuous plant. Sometimes with a solitary stem, but usually having two or three together from the same root, they stand bolt upright from fifteen to twenty-five feet in height. Next to this the commonest conspicuous plant is a large species of *Puya*, belonging to the pine-apple family, with long, stiff, spiky leaves, and these two combined to give a strange and somewhat weird appearance to the vegetation. Here and there were dense masses of evergreen bushes or small shrubs, and more rarely small solitary trees. Among these was probably the species of beech (*Fagus obliqua* of botanists) which the natives call *roble* (or oak), there being, in fact, no native oaks in America south

* This is doubtless the summit described by Darwin under the name Campana de Quillota. He gives the height as 6400 feet above sea-level. The figures in the text are taken from the Chilian survey.

of the equator; but in a passing railway train I could not hope to identify unfamiliar species. Both here and elsewhere in Chili, I noticed that the *quisco* is almost confined to the northern or sunny slopes; while, as Darwin observed, the tall bamboo grass (a *Chusquea*) prevails on the shady sides of the hills.

The summit level, according to Petermann's map, is 4311 feet (1314 metres) above the sea, and thenceforward there is a continuous gradual slope of the ground towards Santiago. The country shows few signs of population, and the larger part of the surface is left in a state of nature, and used only for pasturage in winter. In this arid region cultivation is nearly confined to the valleys of the streams that descend from the Cordillera. The stony beds of the streams passed by the railway were almost completely dried up, and I think that I saw water in one spot only on the whole way between the Aconcagua and the Mapocho.

Any want of interest or variety in the nearer landscape was amply made up by the increasing grandeur of the views of the Cordillera as we approached the capital of Chili, rendered all the more imposing by fresh snow, which extended down to the level of ten or eleven thousand feet. Although it does not include several of the highest summits of the Andes, the range which walls in the province of Santiago to the east is probably the highest continuous portion of the great range; for in a distance of seventy miles, from near the Uspallata Pass to the Volcano of Maipo, I believe that there is but one narrow gap where the crest of the chain falls below the level of nineteen thousand feet.* To the eye, however, the outline seen from the plain is very varied, and by no means gives the impression of a continuous wall. Huge buttresses, with peaked summits, not much inferior in height to the main range, project westward, and in the bays between them form Alpine valleys, which send down streams to fertilize the country. By these buttresses the peak of Tupungato, 20,278 feet in height, the highest summit of this part of the chain, is concealed from Santiago, and I doubt whether it is anywhere visible from the low country on the Chilian side.

Soon after twelve o'clock the train reached the station at Santiago, and I found Mr. Flint, the obliging German proprietor of the Hotel Oddo, in readiness with a carriage to take me to his hotel. The first impression of Santiago, irrespective of the grandeur of its position, is that of a great city. The houses, consisting only of a ground floor, or at most with a single floor overhead, built round an enclosed court, or *patio*, cover a large space, and the town occupies three or four times the area that an equal population would require in Europe. It is laid out, even more regularly than Turin, in square blocks of nearly the same dimensions, so that the ordinary way of reckoning distances is by *quadras*. One enters the

* The mapping of the Andean chain is a task of immense difficulty, and although the Chilian survey is the best that has yet been executed, it leaves much to be desired. Even in the small district which I was able to visit, I found several grave errors in Petermann's map, reduced from the Chilian survey, which is, nevertheless, the best that has been published in Europe. One of the most serious is the omission of the Uspallata Pass, the most frequented of those leading from Central Chili to the Argentine territory, which is neither named nor correctly indicated by the tints adopted to mark the zones of elevation.

town by the Alameda, a straight street, with fine houses on one side and a public garden on the other, nearly two miles in length, along which, at intervals, are statues of the men who have earned the gratitude of their country, the most conspicuous being the equestrian statue of General O'Higgins, the foremost hero of the war of independence.

Turning at right angles into one of the side streets, we soon reached the Hotel Oddo, unpretending in appearance, which was recommended to me as being quieter and more comfortable than the Grand Hotel. This, which was close at hand, occupies the upper floor of a fine pile of building, that fills one side of the Plaza Major, or great square of the city. There seems to be an uneasy feeling that at the first severe shock of earthquake this monument of misplaced architecture may be levelled to the ground, to the destruction of all its inmates.

My first visit in Santiago was made to Don Carlos Swinburne, an English merchant, long established in the city, who has acquired the universal respect and regard of all classes, and whose well-earned personal influence has been on several occasions effective for the mutual benefit of his native land and his adopted country. To his kindness and courtesy I am under many obligations. Later in the day I proceeded to call upon Dr. Philippi, the veteran naturalist, to whom we owe so much of our knowledge of the flora and fauna of Chili.

In Santiago, as in most other South American towns, the first thing that a stranger should do is to learn the routes of the tramcars, which constantly ply through the principal arteries. Hackney coaches are to be found, and are sometimes indispensable, but they are heavy cumbrous vehicles, ill hung on high wheels; one travels slowly and suffers a severe jolting over ill-paved streets. To say nothing of economy, the tramcar runs smoothly at a brisk pace, is usually clean and commodious, and is generally used by all classes of the population. The main point is to take care not to travel in the opposite direction from that intended; but here, with the great landmarks of the Andes always in view, it is not easy to go wrong as to the points of the compass.

To find Dr. Philippi I was directed to a house of modest appearance within the precincts of the Quinta Normal. This establishment is intended to combine the functions of a horticultural garden and a model farm, but the greater part of the grounds appears to be laid out as ornamental pleasure-ground. A large handsome building, originally constructed for a great industrial exhibition, has been turned to good account as a museum of natural history. I was received by Professor Federigo Philippi, who now worthily fills the chair of Natural History in the University of Santiago, from which, after a tenure of many years, his father has retired. Between naturalists none of the ordinary formalities of introduction are required, and cordial relations grow up rapidly. Knowing that Dr. Philippi had already reached an advanced age, I was apprehensive that some infirmity might have chilled the ardour of his interest in science; but I was agreeably disabused when from an adjoining room the professor called his father to join our conversation. I found a man who, although in his seventy-sixth year, was still full of vigour of mind, and I had full opportunity on the following morning to assure myself that this is sustained by abundant physical energy.

Time slips by rapidly in a conversation on subjects of mutual interest, and when, after arranging for a short excursion with Dr. Philippi, I returned homeward, the setting sun was lighting up the heavens with the beautiful tints that are more common in the warm Temperate zone than in other regions of the earth. Low as are the houses, they were just high enough to shut out all but occasional glimpses of the Cordillera from the street; but when I reached the great plaza I came to the conclusion, which I still retain, that Santiago is by many degrees the most beautifully situated town that I have anywhere seen. Rio Janeiro, Constantinople, Palermo, Beyrout, Plymouth, all have the added beauty that the sea confers on land scenery; but such a spectacle as is formed by the majestic semicircle of great peaks that curve round Santiago, lit by the varying tints of day and evening, is scarcely to be matched elsewhere in the world. In position, as in plan of building, I was reminded of Turin; but here the Alps are nearly twice as high, and at half the distance. Further than that, the low country at Turin opens to the east, and, although glorious sunrise effects are not seldom visible, they never rival the splendours of the close of day.

On the following morning, May 12, I started with Dr. Philippi in a hackney coach for an excursion to the Cerro San Cristobal, an isolated hill rising about one thousand feet above the valley of the Mapocho. We crossed that stream by a very massive bridge, constructed to resist the formidable flood poured down the channel after heavy rains, and for about three miles followed the right bank along a rough road deep in the sand formed by the disintegration of the volcanic rocks. We were glad to leave our vehicle at some mills at the foot of the hill, and spent some three hours very agreeably in clambering up and down the rough slopes. The shrubs were much the same as those which I afterwards saw elsewhere in similar situations, but I was fortunate in being introduced to them by one so familiar with the flora as my excellent companion. Among these, as well as the herbaceous plants, the *Compositæ* prevail over every other natural order. Two common species belong to the tribe of *Mutisiaceæ*, unknown in Europe, and almost confined to South America. The bushy species of *Baccharis*, a genus very widely spread in the New World, but not known elsewhere, were also very common. An acacia (*A. Cavenia*) approached more nearly to the dimensions of a tree. It has stiff, spreading, and very spiny branches, and is widely spread throughout the drier parts of temperate South America. Among the few herbaceous plants in flower I was fortunate in seeing the pretty *Gynopleura linearifolia*. This belongs to a tribe of the passion-flower family, very distinct in habit and appearance, which has been by some eminent botanists ranked as a distinct natural order under the name *Malesherbiaceæ*. It includes only two genera with ten or twelve species, all exclusively natives of Chili or Peru.

A veil of morning haze or mist, not uncommon at this season, hung over the city and marred the completeness of the grand view from the summit of the Cerro. Though easily explained, the seeming opacity of a thin stratum of vapour seen from above, as I have often noticed in the Alps, is remarkable. Before we started, and after our return, the haze over the city was scarcely perceptible. Not only did the sun shine brightly in the town, but the outlines of the neighbouring peaks were perfectly distinct. Looking down from the upper station, the slight differences in the intensity of the comparatively feeble light proceeding from the various objects on the surface, by which alone they are made visible, were concealed by the haze which reflected a portion of the comparatively strong light received from the sky, just as when looking from the outside at a window which reflects the light from the sky, we cannot distinguish objects within.

In the afternoon Mr. Swinburne was good enough to accompany me in a visit to Don Benjamin Vicuña Mackenna, one of the most conspicuous and remarkable of the contemporary public men of Chili. His career has been in many ways singular. In early life he took part in two attempts of a revolutionary nature. Fortunately for themselves, the Chilians have gained from their own and their neighbours' experience a fixed aversion to revolution, and, while acknowledging the existence of abuses, have felt that violent change is certain to entail worse evils. Both attempts failed, and the leaders were condemned to death, the sentences being judiciously commuted to temporary exile.

Since his return, Mr. Mackenna has done good

service as head of the municipality of Santiago, has been a prominent member of the legislature, and was, in 1881, the unsuccessful candidate for the presidency of the republic. But it is chiefly by his fertility as a writer that Mr. Mackenna has secured for himself an enduring reputation. Gifted with keen intelligence and a marvellously retentive memory, his readiness to discuss in turn the most varied topics, whether by speech or pen, is quite phenomenal. Besides being a constant contributor to newspapers and periodicals, he has published over a hundred volumes, most of them devoted either to illustrate the history or to promote the progress of his native country. I was most kindly received, and my only regret, on this and subsequent occasions, was that the shortness of my stay prevented me from enjoying more fully the society of this interesting man. From the room—in itself a library—reserved for the spare copies of his own works, I selected four volumes out of the many which he was kind enough to place at my disposal.

On the following day Mr. Vicuña Mackenna was kind enough to devote several hours to taking me to various objects of interest in the city, beginning with the natural history museum at the Quinta Normal. Rightly supposing that they would be of interest, my guide afterwards took me to see the most remarkable trees of the city, each of which possesses some historic interest. In an old and rather neglected garden attached to the palace of the archbishop is the finest known specimen of the peumo, the most important indigenous tree of Central Chili. Popular tradition affirms that under this tree, in 1640, Pedro de Valdivia, the founder of Santiago, held a conference with the native Indian chiefs, in which they agreed to allow the strangers a certain territory for settlement. It is undoubtedly very ancient, and is divided nearly from the ground into a number of massive branches spreading in all directions, so as to form a hemisphere of dark green foliage rather more than sixty feet in diameter. The tree belongs to the laurel family (*Cryptocarya Peumus* of botanists), and is densely covered with thick evergreen leaves impenetrable to the sun. The red oval fruits are much appreciated by the country people, but they have a resinous taste unpalatable to strangers.

In the garden of the Franciscan convent we saw a very fine old Lombardy poplar, from which it is said that all those cultivated in Chili are descended. The story runs that a prior of the convent, who visited his brethren at Mendoza, some time in the seventeenth century, found there poplar trees introduced from Europe, and which in that denuded region were the sole representatives of arboreal vegetation. The sapling which he carried back on his return across the Andes grew to be the tree which still flourishes in the convent at Santiago. To judge from its appearance, the story is no way improbable.

In the *patio* of a fine house in the city are two remarkably fine specimens of the *Eucalyptus globulus*, a tree now familiar to visitors at Nice and many other places in the Mediterranean region. It has been of late extensively planted throughout the drier parts of temperate South America, and promises to be of much economic value. The pair which I saw here had been planted seventeen years before, and, like twins, had kept pace in their growth. The height was about sixty feet, and the girth at five feet from the ground about seven feet.

As a specimen of one of the better houses in Santiago, Mr. V. Mackenna took me to that of one of his cousins, who with his family was at the time absent in the country. The building included three small courts, or patios, each laid out with ornamental plants well watered. The reception-rooms, very richly furnished in satin and velvet, as well as the apartments of the family, were all on the ground floor, most of them opening into a patio. Over a part of the building were small rooms constructed of slight materials for the use of servants, so that the risk of fatal injuries even in a severe earthquake seemed to be but slight.

I was told the history of the owner of this fine house, which, from what I afterwards heard, was no more than a fair sample of the economic condition of Chilian society. Many of the older Spanish families are large landowners, and, in spite of vicissitudes due to droughts and occasional inundations, derive settled incomes from property of this kind. But the prodigious wealth that has flowed from the rich mining districts has proved a temptation too strong to be resisted, and there are comparatively few of the wealthier class who have not engaged in mining speculations. It is needless to say that along with some great prizes there have been many blanks in the lottery, and the result has been that the fortunes of families have undergone the most extraordinary vicissitudes. People get used to a condition of society where the same man may be rich to-day, reduced nearly to pauperism a year later, and then again, after another short interval, rolling in wealth. It is to be feared that the effect, if continued for a generation or two, will not be favourable to progress in the higher sense.

The existence of a class not forced to expend its energies on acquiring wealth, and having some adequate objects of ambition, is still the most important condition for the advancement of the human race. We may look forward to other conditions of society when, having found out the extremely small value of most of the luxuries that now stimulate exertion, men will be able peacefully to develop a healthier and happier social state, in which labour and leisure will be more equally distributed; but this is yet in the distant future, and perhaps the greatest difficulty in its attainment will arise from premature attempts to impose new conditions which, if they are to live, must be of spontaneous growth.

One of the marked features of Santiago is the steep rock of Santa Lucia rising abruptly near the eastern end of the Alameda. It has been well laid out with winding footpaths, and has a frequented restaurant. The view of the snowy range on one side and the city on the other can scarcely be matched elsewhere in the world.

On reaching Santiago, I was mainly preoccupied with the question of how to use my short stay with the best advantage so as to see as much as possible of the scenery and vegetation of the great range, consistently with the promise I had given before leaving home to avoid all risks to health. From the abundance of fresh snow along the range, it was obvious that the precipitation on the higher flanks of the Cordillera must be considerably greater than it is in the low country, where only one or two slight showers had fallen; and we were in the season when rain is annually expected, which, of course, would take the form of snow in the higher region. I had already obtained a letter to the manager of the mines at Las Condes, a place about fifteen miles from Santiago, and some eight thousand feet above the sea. But, after taking counsel with those best informed, I decided on giving a few days to a visit to the Baths of Cauquenes, in the valley of the Cachapoal, a little above the point where that stream issues from the mountains into the plain of Central Chili. There remained a possibility of making an excursion from Cauquenes into one of the interior valleys, especially that of Cypres, famed for the variety of high mountain plants that find a home near the glacier which descends into it, and there was the advantage that even in case of bad weather no serious inconvenience would arise.

I started next morning, May 14, by the railway, which is carried nearly due south from the capital to Talca, and thence to Concepcion. I found myself in the same carriage with Mr. Hess, the lessee and manager of the Baths, an energetic, practical man, fully impressed with a sense of his own importance as head of an establishment which annually attracts the best society of Chili. The railway journey, which carries one for about fifty miles parallel to the great range of the Cordillera, is very interesting, even at this season, when much of the country shows a parched surface. The finest views are those gained where the line passes opposite the opening through which the Maipo issues from the mountains into the plain. This river, which even in the dry season shows a respectable volume of water, is formed by the union of the torrents from four valleys that penetrate nearly to the axis of the Cordillera. Of Tupungato, the highest summit hereabouts, 20,270 feet above sea-level, I saw nothing, as it is masked by a very lofty range that divides two of the tributary valleys. A slender wreath of vapour marked the volcano of San José, just twenty thousand feet in height, at the head of the southern branch of the river. It is only at one point visible from the railway.

On the way from Valparaiso to Santiago I had already been much struck by the prevalence over wide areas of plants not indigenous to the country, most of them introduced from Southern Europe. The most conspicuous are plants of the thistle tribe, all strangers to South America, and especially the cardoon, or wild state of the common artichoke. This is now far more common in temperate South America than it anywhere is in its native home in the Mediterranean region. In Chili it is regarded with some favour, as mules, and even horses, eat the large spiny leaves freely at a season when other forage is scarce. The same cannot be said of our common coarse spear-thistle (*Cnicus lanceolatus*), which, though of much more recent introduction, has now invaded large tracts of country, especially in the rather moister southern provinces. I was informed that, with the strange expectation that it would be useful as fodder, an Englishman had imported a sack of the seed, which he had spread broadcast somewhere in the neighbourhood of Concepcion. Many other European plants have been introduced, either intentionally or by accident, and have in some districts to a great extent supplanted the indigenous vegetation. As to many of these, it appears to me probable that their diffusion is due more to the aid of animals than the direct intervention of man. This is especially true of the little immigrant which has gone farthest in colonizing this part of the earth—the common stork's-bill (*Erodium cicutarium*), which has made itself equally at home in the upper zone of the Peruvian Andes, in the low country of Central Chili, and in the plains of Northern Patagonia. Its extension seems to keep pace with the spread of domestic animals, and, as far as I have been able to ascertain, it is nowhere common except in districts now or formerly pastured by horned cattle. It is singular that the same plant should have failed to extend itself in North America, being apparently confined to a few localities. It is now common in the northern island of New Zealand, but has not extended to South Africa, where two other European species of the same genus are established.

In considering the facts relating to the rapid extension of certain plants when introduced into new regions, and the extent to which they have supplanted the indigenous species, I confess that I have always been a little sceptical as to the primary importance attributed by Darwin* to the fact that most of these invaders are northern continental species. In the course of a long existence extending over wide areas, he maintains that these have acquired an organization fitting them better to maintain the struggle for existence than the indigenous species of the regions over which they have spread. Of course, it is true in the case of territories very recently raised from the sea, and not in direct connection with a continental area inhabited by species well adapted to the conditions of soil and climate, that immigrant species well adapted to the conditions of their new home will spread very rapidly, and may easily supplant the less vigorous, because less well adapted, native species. The most remarkable case of this kind is perhaps presented by Northern Patagonia and a portion of the Argentine region, raised from the sea during the most recent geological period. The only quarters from which the flora could be recruited were the range of the Andes to the west, and the subtropical zone of South America to the north. Everything goes to prove that the forms of plants are far more slowly modified than those of animals—or, at least, of the higher vertebrate orders. The new settlers are unable quickly to adapt themselves to the new conditions of life, and as a result we find that the indigenous flora of the region in question is both numerically poor in species, and that these have been unable fully to occupy the ground. Among the species intentionally or accidentally introduced by the European conquerors, those well adapted

---

* "Origin of Species," 3rd edit., p. 410.

to the new country have established a predominance over the native species; but I question whether, if the course of history had been different, and the conquerors of South America had come from South Africa or South Australia, bringing with them seeds of those regions, we should not have seen in Patagonia African or Australian plants in the place of the European thistles and other weeds now so widely spread.

If I am not much mistaken as to the history of the introduction of foreign plants into new regions, it very commonly happens that a species which spreads very widely at first becomes gradually restricted in its area, and finally loses the predominance which it seemed to have established. Attention has not, I think, been sufficiently directed to the fact that the chief limit to the spread of each species is fixed by the prevalence of the enemies to which it is exposed, and, that plants carried to a distant region will, as a general rule, enjoy advantages which in the course of time they are likely to lose. Whether it be large animals that eat down the stem—as goats prevent the extension of pines— or birds that devour the fruit, or insects that attack some vital organ, or vegetable parasites that disorganize the tissues, the chances are great that in a new region the species will not find the enemies that have been adapted to check its extension in its native home. Of the marvellous complexity of the agencies that interact in the life-history of each species we first formed some estimate through the teachings of Darwin; but to follow out the details in each case will be the work of successive generations of naturalists. We cannot doubt that in a new region new enemies will arise for each species that has become common, or, in other words, that other organisms, whether animals or plants, will acquire the means of maintaining their own existence at the expense of the new-comer. The wild artichoke is doubtless perfectly adapted to the climate of the warmer and drier parts of the Mediterranean region, and is there rather widely spread ; but it is nowhere very common, even in places where the ground is not much occupied by other species. We do not know all the agencies that prevent it from spreading farther, but we do not doubt that it is held in check by its appropriate enemies. In South America it would appear that these, or some of them, are absent, and the plant has spread far and wide. If some common bird should take to devouring the seeds, or some other effectual check should arise, the area would very speedily be reduced.

The train stopped for breakfast at the Rancagua station, a few miles from the town of that name. Along with very fair food at the restaurant, cheaper delicacies were offered by itinerant hawkers, including various sweet cakes of suspicious appearance and baskets of red berries of the *peumo* tree. At the next station, called Gualtro, about fifty miles from Santiago, we left the train, and, after the usual long delay, continued our journey in a lumbering coach set upon very high wheels. This seems to be the general fashion for carriages in South America, arising from the fact that the smaller streams, which swell fast after rain, are usually unprovided with bridges.

Incautious travellers in South America may easily be misled by the frequent use of the same name for quite different places. One bound for the Baths of Cauquenes must be careful not to confound these with the town of Cauquenes, the chief place of a department of the same name, more than a hundred miles farther to the south.

Before reaching Gualtro we had crossed the Cachapoal, a torrential stream which drains several valleys of the high Cordillera. Our course now lay eastward, towards the point where the river issues from the mountains into the plain, and where, as everywhere in Central Chili, its waters are largely used for irrigation. The road along the left bank lies on a slope at some height above the stream, and gives a wide view over the plain, backed by the great range of the Cordillera. Irrespective of the picturesque interest of the grand view, I added somewhat to the impressions respecting the physical geography of Central Chili which I had recently received from an examination of Petermann's reduction from the large government map, and from the information given me at Santiago.

I had reached Chili with no other ideas respecting the configuration of the country than those derived from the twelfth chapter of Darwin's "Journal of Researches," which with little modification have been repeated by subsequent writers, even so lately as in the excellent article on "Chile," in the American Cyclopædia.

Struck by the conformation of the range between Quillota and Santiago, and the somewhat similar range south of the Maipo, and writing at a time when there were no maps deserving of the name, and when the channels of Patagonia had been most imperfectly explored, Darwin was led to infer a much closer resemblance between the orographic features of the two regions than it is now possible to admit. He supposed the greater part, if not the whole of the Chilian coast, to be bordered by mountain ranges running parallel to the main chain of the Cordillera, thus forming a succession of nearly level basins lying between these outer mountains and the main range, each being drained through a tranverse valley which cuts through the outer range. Such a conformation of the surface would undoubtedly resemble what we find on the western coast of South America, between the Gulf of Ancud and the Straits of Magellan. But the facts correspond with this view only to a limited extent.

The tints laid down on Petermann's map to indicate successive zones of height above the sea are far from being completely accurate, but slight errors of detail do not affect the general conclusions to which we must arrive. If we carry the eye along from north to south, we find a succession of great buttresses or promontories of high land projecting westward from the main range, between which relatively deep valleys carry the drainage towards the Pacific coast. The effect of a continuous sinking of the land would be to produce a series of deep bays running far inland to the base of the Cordillera, and further depression might show here and there some scattered islets, but nothing to resemble the almost continuous range of mountainous islands that separate the channels of Patagonia from the ocean. As far as it is possible to judge of a region yet imperfectly surveyed, the case is quite different in Southern Chili, below the parallel of 40°. From near Valdivia a lofty coast range, cut through by only one deep and narrow valley, extends southward to the strait, only a few miles wide, that divides the island of Chiloe from the mainland, and is evidently prolonged to the southward in the high land that fringes the western flank of that large island. A moderate rise of the sea-level would submerge the country between Puerto Montt and the Rio de San Pedro, and produce another island very similar in form and dimensions to that of Chiloe.

Our lumbering carriage came to a halt at the place where the road crosses a stream—the Rio Claro—which drains some part of the outer range and soon falls into the Cachapoal. Close at hand was a plain building with numerous dependencies, which turned out to be the residence of Don Olegario Soto, the chief proprietor of this part of the country. I proceeded at once to deliver a letter to this gentleman, whose property extends along the valley for a distance of thirty or forty miles into the heart of the Cordillera. My object was to ascertain the possibility of making an excursion into the interior of the great range, and to obtain such assistance as the proprietor might afford. The house, so far as I saw, was rustic in character, and my first impression of its owner was that the same epithet might serve as his description. There was a complete absence of the conventional and perfectly hollow phrases which form the staple of Castilian courtesy. But first impressions are proverbially misleading. On my making some obviously superfluous remark as to my imperfect use of the Spanish tongue, Don Olegario changed the conversation to English, which he spoke with perfect ease and correctness. We discussed my project of a mountain excursion, and I found at once that he was ready to give practical assistance in every way. The doubt remained as to the season and the weather. If no rain or snow should fall, there was no other obstacle. He readily undertook to provide men and horses and everything needful for an excursion of three days in the Cordillera, and I was to let him know my resolve on the following day.

I afterwards heard in some detail the family history of this liberal-minded gentleman. His father commenced life as a common miner. With the aid of good fortune, natural intelligence, and activity, he became the owner of a valuable mine in Northern Chili, and amassed a large fortune, mainly invested in the purchase of land. Having several sons, he sent them all for education to England, and, to judge from the specimen I saw, with excellent results. Large proprietors who use intelligence and capital to develop the natural resources of the country supply, in some states of society, the most effectual means for progress in civilization ; but, excepting in Chili, such examples are rare in South America.

The day was declining when we reached the Baths of Cauquenes, and I had time only for a short stroll through the establishment and its immediate surroundings. It stands on a level shelf of stony ground less than a hundred feet above the river Cachapoal, the main building consisting of a range of bedrooms, all on the ground floor, disposed round a very large quadrangle. The rooms are spacious and sufficiently furnished, and I was struck by the fact that there is no fastening whatever to the doors, which usually stand ajar. This speaks at once for the constant apprehension of earthquakes that seems to haunt the Chilian mind, and for the general honesty of the people, amongst whom theft is almost unknown. Besides some additional rooms in wings adjoining the great court, the baths are an *annexe* overhanging the river, to which you descend by broad flights of stairs. A large handsome hall, lighted from above, has the bath-rooms ranged on either side, all exquisitely clean and attractive. The adjoining ground, planted mainly with native trees, is limited in extent. A narrow and deep ravine, cut through the rocky slope of the adjoining hill, is traversed by one of those slight wire suspension bridges common in this country, that swing so far under the steps of the passenger as to disquiet the unaccustomed stranger. The views gained from below up the rugged and stern valley of the Cachapoal are naturally limited, but the rather steep hills rising above the baths promised a wider prospect towards the great range of the Cordillera, and did not disappoint expectation.

The autumn season being now far advanced, the guests at the establishment were few—about twenty in all. After supper they assembled in a drawing-room and adjoining music-room. I was struck not only by the general tone of courtesy and good-breeding of the party, but by the fact that several of them at least were well-informed men, taking an intelligent interest in physics and natural history. Two or three gentlemen spoke a little, but only a little, English, and, my command of Spanish being equally imperfect, conversation did not flow very freely, and I retired for the night with a feeling that at a more favourable season I should be very loth to quit such pleasant headquarters.

After a rather cold night, I rose early on the 15th of May, with a sense of the impending necessity for an immediate decision as to my future plans. Scanning anxiously the portion of the great range seen towards the head of the valley, I saw that fresh snow extended much lower than I had observed it at Santiago, while heavy broken masses of dark clouds lay along the flanks of the higher mountains. I received no encouragement from Mr. Hess. The ordinary season for rain in the low country had arrived, and this would take the form of snow in the inner valleys of the Cordillera; all appearances boded a change of weather which is always anxiously desired by the native population. I reluctantly decided to despatch a messenger to Don Olegario Soto renouncing the projected excursion, contenting myself with the prospect of approaching as near to the great range as could be accomplished in a single day from the baths.

To the naturalist, however, a new country is never devoid of interest; and this was my first day on the outer slopes of the Chilian Andes. The season was, indeed, the most unfavourable to the botanist of the entire year. After six months' drought, broken only by one or two slight showers, the ground was baked hard, nearly to the consistence of brick, and most of the herbaceous vegetation utterly dried up. A great part of the day was nevertheless very well spent in rambling over the hill above the baths, and making closer acquaintance with many vegetable forms altogether new, or hitherto seen only from a distance. The trees and shrubs of this region are with scarce an exception evergreen, and the most conspicuous, though differing much from each other in structure and affinities, bear a striking resemblance in the general form and character of their foliage, formed of thickset, broadly elliptical, leathery leaves, giving a dense shade impervious to the sun. The largest is the *peumo** tree, already referred to, which forms a thick trunk, but rarely exceeds thirty feet in height. Next to this in dimensions are two trees of the Rosaceous family, allied in essential characters (though very different in appearance) to the Spiræas, of which the common meadowsweet is the most familiar example. One of these, the *Quillaja saponaria* of botanists, is much prized for the remarkable properties of the bark, said to contain, along with carbonate of lime and other mineral constituents, much *saponine*, an organic compound having many of the properties of soap. It is commonly used for washing linen, and especially for cleansing woollen garments, to which it gives an agreeable lustre. Nearly allied to this is the *Kageneckia oblonga*, a small tree of no

---

* Molina, one of the most pernicious blunderers who have brought confusion into natural history, grouped together under the generic name *Peumus* several Chilian plants having no natural connection with each other. Misled by his erroneous description, botanists have applied the name *peumus* to a fragrant shrub, common about Valparaiso and elsewhere, which is known in the country by the name *boldu*.

special use except to aid in clothing the parched hills of the lower region of Chili. It would seem that all these trees might be successfully introduced into the warmer parts of southern Europe, especially the south of Spain and Sicily, and the *Quillaja* would doubtless prove to be of some economic value.

To the European traveller the most remarkable vegetable inhabitant of the dry hills of Central Chili is the tall cactus (*Cereus quisco*), which I had first seen on the way from Valparaiso to Santiago. They were abundant on the lower slopes about Cauquenes, the stiff columnar stems averaging about a foot in diameter. I was told that the plant was now to be found in flower, and was surprised to observe on the trunks, as I approached, clusters of small deep-red flowers that appeared very unlike anything belonging to this natural family. Nearer inspection showed that they had none but an accidental connection with the plant on which they grew. The genus *Loranthus*, allied to our common European parasite, the mistletoe, is widely spread throughout the world, chiefly in the tropics. From three to four hundred different species are known, nearly all parasites on other plants; as a rule, each species being confined to some special group, and many of them known to fix itself only upon a single species. Botanists in various regions have remarked that there is frequently a marked resemblance between the foliage of the parasitic *Loranthus* and that of the plants to which it is attached; but it is especially remarkable that the only species which is known to grow upon the leafless plants of the cactus family should itself be the only

leafless species of *Loranthus*, consisting as it does
only of a very short stem, from which the crowded
flower-stalks form a dense cluster of bright-red,
moderately large flowers. Although it is not easy to
conjecture how it may act, it is conceivable that these
conformities may be results of natural selection ; but
it is also possible that, like many curious instances of
parallelism among the forms of plants belonging to
widely different types, the facts may hereafter be seen
to result from some yet undiscovered law regulating
the direction of variation in the development of
organic beings.

In some places dense masses of spiny shrubs were
massed together, overgrown by climbing plants,
amongst which the most strange and attractive were
composites of the genus *Mutisia*. The Chilian species
have all stiff, leathery, undivided leaves ending in a
tendril, with large brownish-red or purple flowers, of
which very few were to be found at this advanced
season. Among the shrubs I was struck by a species
of *Colletia*, a genus characteristic of temperate South
America. They are nearly or quite leafless, and
remind one slightly of our European furze, but are
much more rigid, with fewer, but hard and penetrating
spines, which, unlike those of the furze, are true
branches, sharpened to a point and set on at right
angles to the stem. The species common here
(*Colletia spinosa* of Lamarck) grows to a height of
four or five feet, and would probably be found very
useful for hedges on dry stony ground in the south of
Europe. I regret that the seeds which I sent to Italy
have not germinated.

At the present season, corresponding to mid-November in Europe, I could not expect to see much of the native herbaceous vegetation, and the majority of the plants collected showed little more than the parched skeletons of their former selves. The recent slight showers, which alone had broken the long drought since the preceding spring, sufficed to awaken into life two species of *Oxalis*, whose flowers and early leaves just pierced through the hard surface of the soil; but, although some young leaves heralded the appearance of species of the lily tribe, no other new flowers had appeared. Ferns were scarce, but I was rather surprised to find a fine *Adiantum* in some abundance under the shade of the *Quillaja* and *Kageneckia* trees.

In the evening I arranged with Mr. Hess to start early on the following morning, with the object of approaching as nearly as possible to the higher zone of the Cordillera, of which, despite cloudy weather, I had tempting glimpses during the day.

I was on foot early on the 16th, but the prospect was not altogether cheering. The clouds which covered the sky were of leaden hue, and lay about mid-height on the range of the Cordillera. The horses were ready after the usual delay, and a taciturn young man, who probably thought the expedition a bore, was in readiness to act as guide. As I was about to mount, Mr. Hess lent me a *poncho*, which I at once drew over my head, and for which I afterwards had reason to be grateful. We rode on in silence for more than an hour, following a track that cuts across the great bend of the Cachapoal above the baths. The river is formed by the union of four or five torrents that issue from as many of the interior valleys of the Cordillera. It flows at first northward, nearly parallel to the main chain, until, a few miles above the baths, it bends westward and descends towards the open country. We had reached a point overlooking the upper valley, and, as far as one might judge from glimpses through breaks in the clouds, commanding a noble view of the great range of the Cordillera. Before us lay the slopes by which, at a distance of two or three miles, we might reach the only bridge which spans the upper course of the Cachapoal. Just at this interesting point the threatened rain began, at first gentle, but steadily increasing. I went on for some time on the chance of any token of improvement; but, as none appeared, I decided on sending back the horses and returning on foot to the baths.

I had this day my first experience of the value of a genuine *poncho* woven by the Indian women from the wool of the guanaco. Throughout South America the cheap articles in common use, manufactured in England and Germany, have almost replaced the native garment. They are comparatively heavy and inconveniently warm, while not at all efficient in keeping out rain. After more than three hours' exposure to heavy rain, the light covering lent to me by Mr. Hess had allowed none to pass. It is surprising that such a serviceable and convenient garment, which leaves the arms free, and is equally useful on foot or on horseback, is not more generally adopted in Europe, especially by sportsmen. A good *poncho* is not, however, to be had cheaply. I was asked sixty dollars for one at Buenos Ayres, and that, I believe, is about the ordinary price.

The change of weather which culminated in this wet day at Cauquenes seems to have extended along the range of the Cordillera; but, to illustrate the rapid change of climate which is found in advancing northward along the west side of the Andes, I may mention that, while the rain continued to fall steadily for ten and a half hours at Cauquenes, it lasted but five hours at Santiago, about fifty miles to the northward; and at Santa Rosa, forty miles farther in a direct line, only two hours' rain was obtained by the thirsty farmers on the banks of the Aconcagua.

On the morning of the 17th the clouds had disappeared, and the valley was lit up with brilliant sunshine. Fresh snow lay thickly on the flanks of the higher mountains, and I had reason to congratulate myself that I had not undertaken an expedition which would have resulted in utter discomfort without any adequate compensation, as the Alpine vegetation must have been completely concealed by the fresh snow. The roads and paths were all deep in mud, and the slopes very slippery from the rain, so I decided on descending to the rocky banks of the river below the baths, and, following the stream as far as I conveniently could. I did not go far, but a good many hours were very well occupied in examining the vegetation of the left bank of the Cachapoal and of a little island of rock in the middle of the stream. In summer one of the ordinary suspension bridges of the country enables the visitors to cross to the right bank, but this is removed during winter, and the swollen waters of the river made all the usual fords impassable for the present.

Many forms of *Escallonia* were abundant along the stream. A few species only of this genus are cultivated in English gardens, but in their native home, the middle and lower slopes of the Andes, they exhibit a surprising variety of form while preserving a general similarity of aspect. They are all evergreen shrubs, some rising to the stature of small trees, with undivided, thick, usually glossy leaves, and white, red, or purplish flowers. Although forty-three different species have been described from Chili alone, it is easy to find specimens not exactly agreeing with any of them, and to light upon intermediate forms that seem to connect what appeared to be quite distinct species. They afford an example of a fact which I believe must be distinctly recognized by writers on systematic botany—that in the various regions of the earth there are some groups of vegetable forms in which the processes by what we call species are segregated are yet incomplete; and amid the throng of closely allied forms, the suppression of those least adapted to the conditions of life has not advanced far enough to differentiate those which can be defined and marked by a specific name.

To the believer in evolution, it must be evident that at some period in the history of each generic group there must have occurred an interval during which species, as we understand them, did not yet exist; and perhaps the real difficulty is to explain why such instances are not more frequent than they now appear to be. Familiar examples are the genera *Hieracium* and *Rosa* in Europe; *Aster* and *Solidago* in North America; while in South America, *Escallonia*, *Malvastrum*, and several groups of *Myrtaceæ* seem to exhibit the same phenomenon.

Another genus having numerous species in South America, but, so far as I know, not displaying the same close connection of forms linking the several species, is *Adesmia*, a leguminous genus allied to the common sainfoin. I found several species near the baths, the most attractive being a little spiny yellow-flowered bush, with much the habit of some Mediterranean *Genistæ*, but with pods formed of several joints, each covered with long, purple, glistening hairs.

A bright day was followed by a clear cold night, the thermometer falling to 40° Fahr. in the court, and slight hoar-frost was visible in the lower part of the valley near the baths. I started early for a ramble over the higher hills rising to the south and southwest of the establishment. After following a track some way, I struck up the steep stony slopes, meeting at every step the dried skeletons of many interesting plants characteristic of this region of America, but here and there rewarded by finding some species in fruit, or even with remains of flower. After gaining the ridge, I found that the true summit lay a considerable way back, quite out of sight of the baths. To this, which is called *El Morro de Cauquenes*,[*] I directed my steps, wishing to enjoy a unique opportunity for a wide view of the Chilian Andes.

---

[*] The Baths of Cauquenes are said to be 2523 feet above the sea; the *Morro*, by aneroid observation, is about 2000 feet higher.

The day was cloudless, and the position most favourable. In this part of the range the Cordillera bends in a curve convex to the east, so as to describe a nearly circular arc of about 60°, with Cauquenes as a centre. The summits of the main range, which apparently vary from about sixteen to nineteen thousand feet in height, and are nearly forty miles distant, send out huge buttresses dividing the narrow valleys whose waters unite to form the Cachapoal, and are in many places so high as to conceal the main range. The slopes are everywhere very steep, so that, in spite of the recent fall of snow, dark masses of volcanic rock stood out against the brilliant white that mantled the great chain. The tints in Petermann's map would indicate that the highest peaks are those lying about due east, but it appeared to me that two or three of those which I descried to the south-east, though slightly more distant, were decidedly higher. It will probably be long before the Chilian Government can undertake a complete survey of the gigantic chain which walls in their country on the eastern side. No pass, as I was informed, is used to connect the upper valley of the Cachapoal with the Argentine territory.

From the summit I descended about due north into a little hollow, whence a trickling streamlet fell rather rapidly towards the main valley. As commonly happens in Chili, this has cut a deep trench, or *quebrada ;* and when I had occasion to cross to the opposite bank, I had no slight difficulty in scrambling down the nearly vertical wall, though partly helped and partly impeded by the shrubs that always haunt these favourable stations. The Winter's bark, not yet in flower, differed a good deal from the form which I had seen at Valparaiso, and the foliage was much the same that I afterwards found in the channels of Patagonia. Among the few plants yet flowering at this season was a large lobelia, of the group formerly classed as a distinct genus under the name *Tupa*,* and which is peculiar to Chili and Peru.

* As happens with many other plants described by early botanists, there has been much confusion in regard to the species named by Linnæus *Lobelia Tupa*. The plant was first made known to Europeans by the excellent traveller, Father Feuillée, whose "Journal des Observations Physiques Mathématiques et Botaniques faites sur les côtes de l'Amérique meridionale, etc.," published in 1714, is a book which may still be consulted with advantage. His descriptions of plants are usually careful and accurate, but the accompanying plates all ill-executed and often misleading. Linnæus, followed by Willdenow, refers to Feuillée's work, but gives a very brief descriptive phrase which suits equally well Feuillée's plant and several others subsequently discovered. Aiton, in the "Hortus Kewensis," gives the name *Lobelia Tupa* to a plant which is plentiful about Valparaiso, where I found it still in flower, the seeds of which were received at Kew about a century ago from Menzies. This is now generally known by the not very appropriate name *Tupa salicifolia* of Don, but was first published by Sims in the *Botanical Magazine*, No. 1325, as *Lobelia gigantea*, which name it should now bear. The plant which I found near Cauquenes appears to be the *Tupa Berterii* of Decaudolle, a rare species, apparently not known to the authors of the "Flora Chilena." No doubt could have arisen as to the plant intended by Linnæus as *Lobelia Tupa* if writers had referred to Feuillée's full and accurate description. His account of the poisonous effects of the plant was probably derived from the Indians, and may be exaggerated. The whole plant, he says, is most poisonous, the mere smell causing vomiting, and any one touching his eyes after handling the leaves is seized with blindness. I may remark that the latter statement, which appears highly improbable, receives some confirmation from the observations of Mr. Nation, mentioned above in page 77. The plant which I saw in Peru, but failed to collect, is much smaller than most of the Chilian species, and has purple flowers, but is nearly allied in structure. It is probably the *Tupa secunda* of Don. I gather from a passage in one of Mr. Philippi's writings that the word *tupa* in Araucanian signifies poison. We are yet, I believe, ignorant of the

On my return from a delightful walk, I found much-desired letters from home awaiting me, and along with them the less welcome information that the departure of the *Triumph* was delayed for several weeks. Renouncing with regret the agreeable prospect of a voyage in company with Captain Markham, I at once wrote to secure a passage in the German steamer *Rhamses*, announced to leave Valparaiso on May 28.

Among other objects of interest at this place, I was struck by the proceedings of two captive condors, who, with clipped wings, roamed about the establishment, and seemed to have no desire to recover the liberty which they had lost as young birds. One of them was especially pertinacious in keeping to the side of the court near to the dining-room and kitchen, always on the look-out for scraps of meat and refuse. Contrary to my expectation, the colour of both birds, which were females, was a nearly uniform brown, with only a few white feathers beneath. They were larger than any eagles, but scarcely exceeded one or two of the largest *lämmergeier* of the Alps that I have seen in confinement.

On the morning of May 19 I with much regret took my departure from the baths, and found myself in company with an elderly gentleman and his pretty and agreeable daughter, who also desired to return to Santiago. Starting some two hours earlier than was at all necessary, we had spare time, which I employed in looking for plants at Rio Claro and about the

chemical nature of the poisonous principle contained in the plants of this group.

Gualtro station; but at this season very little remained to interest the botanist. We reached the capital about five p.m., and, as the days were now short, the sun was setting as I went in an open carriage along the broad Alameda, which runs nearly due east. The better to enjoy the finest sunset which I had yet seen in America, I was sitting facing westward, with my back to the horses, when an unusual glow of bright light on the adjoining houses caused me to turn my head. Never shall I forget the extraordinary spectacle that met my eyes. I am well used to brilliant sunsets, for, so far as I know, they are nowhere in the world so frequent as in the part of north-eastern Italy approaching the foot of the Alps, with which I am familiar. But the scene on this evening was beyond all previous experience or imagination. The great range of the Cordillera that rises above the town, mostly covered with fresh snow, seemed ablaze in a glory of red flame of indescribable intensity, and the whole city was for some minutes transfigured in the splendour of the illumination.

The subject of sunset illumination has been much discussed of late in connection with the supposed effects of the great eruption of Krakatoa, and I confess to a suspicion that these have been considerably overrated. That the presence of finely comminuted particles in the higher region of the atmosphere is one of the chief causes that determine the colour of the sky, may be freely conceded by those who doubt whether a single volcanic eruption sufficed to alter the conditions over the larger part of the earth's surface. It is certain that some of the districts ordinarily noted for sunsets of extraordinary brilliancy are remote from active volcanoes. So far as South America is concerned, it may, on the other hand, be remarked that if volcanic action be an efficient cause, it is present at many points of the continent as well as in Central America, while brilliant sunsets are, so far as I know, of rare occurrence except in Chili.

CHAPTER IV.

Baths of Apoquinto—Slopes of the Cordillera—Excursion to Santa Rosa de los Andes and the valley of Aconcagua—Return to Valparaiso—Voyage in the German steamer *Rhamses*—Visit to Lota—Parque of Lota—Coast of Southern Chili—Gulf of Peñas—Hale Cove—Messier's Channel—Beautiful scenery—The English narrows—Eden harbour—Winter vegetation—Eyre Sound—Floating ice—Sarmiento Channel—Puerto Bueno—Smyth's Channel—Entrance to the Straits of Magellan—Glorious morning—Borya Bay—Mount Sarmiento—Arrival at Sandy Point.

HAVING devoted the day following my return to Santiago to botanical work, chiefly in the herbarium of Dr. Philippi, I started on the following morning in company with his son, Professor Friedrich Philippi, for an excursion up the slopes of the mountain range nearest the city. My companion had kindly sent forward in advance his servant with horses, and we engaged a hackney coach to convey us to the Baths of Apoquinto, where a warm mineral spring bursts out at the very base of the mountain. The common carriages throughout South America are heavy lumbering vehicles, and the road, though nearly level, was deep in volcanic sand; but the horses are excellent, and, in spite of several halts to collect a few plants yet in flower, we accomplished the distance of nine miles in little over an hour.

The establishment at Apoquinto is on a small scale and somewhat rustic in character, but it had been recently taken by an Englishman, and now supplies fair accommodation, which would be prized by a naturalist who should be fortunate enough to visit Chili at a favourable season. We mounted our horses without delay, and at once commenced the ascent, gentle for a short way, but soon becoming so steep that it was more convenient to dismount at several places. Under the experienced guidance of my companion, I found more interesting plants still in flower or fruit than I had ventured to expect at this season. I here for the first time found a species of *Mulinum*, one of a large group of umbelliferous plants characteristic of the Chilian flora, and nearly all confined to South America. The leaves in the commonest species are divided into a few stiff pointed segments, reminding one somewhat of the *Echinophora* of the Mediterranean shores, once erroneously supposed to be a native of England.

I was especially struck on this day with the extraordinary variety of odours, pleasant or the reverse, that are exhaled by the native plants of Chili. As commonly happens in dry countries, a large proportion of the native plants contain resinous gums, each of which emits some peculiar and penetrating smell. I had already observed this elsewhere in the country, but, perhaps owing to the great variety of the vegetation on these slopes, the recollections of the day are indelibly associated with those of the impressions on the olfactory nerve. If there be persons in whom such impressions are sufficiently distinct to be accurately recalled by an effort of the memory, I can imagine that in some countries the nose might afford a valuable help to the botanical collector. To judge, however, from personal experience, I should say that of all the senses that of smell is the one which supplies the least accurate impressions, and those least capable of certain recognition.

We reached a place where a small stream from the upper part of the mountain springs in a little waterfall from a cleft in the rocks, and which is known as the Salto de San Ramon. This is probably about four thousand feet above the sea-level, and between us and the lower limit of the snow which covered the higher slopes there stretched a rather steep acclivity, covered, like the ground around us, with bushes and small shrubby plants. A few small trees (chiefly *Kageneckia*) grew near the Salto, but higher up scarce any were to be seen. Professor Philippi, who is well acquainted with the ground, thought that little, if anything, would be added to our collections by continuing the ascent, so we devoted the spare time to examining the ground in our immediate neighbourhood, thus adding a few species not before seen. In summer, however, an active botanist, starting early from Apoquinto, who did not object to an ascent of six or seven thousand feet, would reach the zone of Alpine vegetation, and be sure to collect many of the curious plants of this region of the Andes.

May 22 and the following day were fully occupied in Santiago. Among other agreeable acquaintances,

I called upon Don F. Balmacedo, then minister for foreign affairs, and now President of the Republic, who favoured me with a letter of introduction to the governor of the Chilian settlement in the Straits of Magellan. I also enjoyed an interesting conversation with Dr. Taforò, then designated by the Chilian Government for the vacant archbishopric of Santiago. Some canonical objections appear to have created difficulties at Rome, and the see, as I believe, remains vacant.

I found in Dr. Taforò an agreeable and well-informed gentleman, who appeared to hold enlightened views, and to be free from many of the prejudices which the Spanish clergy have inherited from the dark period of ecclesiastical tyranny and absolute royalty. With regard to the Chilian clergy in general, I derived a favourable impression from the testimony of my various acquaintances. At all events, they appear to be respected by the mass of the population, whereas in Peru they are regarded with dislike and contempt by all classes alike.

Among the various claims of the Chilian republic to be regarded with interest by the student of political progress, I must note the fact that it has for some time successfully adopted a system of suffrage which is supposed to be too complex for the people of our country. In political elections for representatives the mode of voting is, I believe, very nearly the same as that known amongst us as the Hare system; while in municipal elections the cumulative vote is adopted, each voter having as many votes as there are candidates to be elected, and being allowed to give as many votes as he pleases to the one or more candidates of his choice. I unfortunately was not aware of these facts while in the country, and therefore failed to make inquiry on the subject; but the fact that, while there is a keen interest in political life, no one has proposed to alter the present mode of voting, seems to prove that the existing system gives general satisfaction.

Early in the morning of May 24 I left Santiago, bound for Santa Rosa de los Andes, the highest town in the valley of the Rio Aconcagua. That river is mainly fed from the snows of the great peak from which it takes its name, the highest summit of the New World.* In its lower course it waters the Quillota valley, through which the railway is carried from Valparaiso to Santiago. In travelling from the latter city it is therefore necessary to return to the junction at Llaillai, whence a branch line leads eastward along the river to San Felipe and Santa Rosa. The sky was cloudless, the air delightfully clear, and the views of the great range were indescribably grand and beautiful, especially in the neighbourhood of San Felipe. The summit of Aconcagua, as seen from this side, shows three sharp peaks of bare rock, too steep to retain the snow which now lay deep on the lower declivities. It has been inferred that the summit must be formed of crystalline or metamorphic rock, as there is no indication of the existence of a crater. This is by no means improbable, as we know that

---

* The measurements of the height of the peak of Aconcagua vary considerably in amount, but I believe that the most reliable is that adopted by Petermann—6834 metres, or 22,422 English feet.

granite, old slates, and conglomerates, as well as newer Secondary rocks, are found at many points along the axis of the main range ; but, on the other hand, we know that most of the higher peaks in Central Chili are volcanic, and the removal of all but some fragments of the cone of an ancient crater may leave sharp teeth of rocks such as are seen at the summit of Aconcagua. In the view which I obtained from the Morro of Cauquenes I observed several lofty peaks of somewhat the same character, which struck me as probably the shattered remains of ancient craters.

Reaching Santa Rosa early in the afternoon, I proceeded to the Hotel Colon in the *plaza*, which, as usual, forms the centre of the town. The French landlord and his wife were civil, obliging people, and, although the establishment seemed to be much out at elbows, I was soon installed in a tolerably good room, and supplied with information for which I had hitherto been vainly seeking. The main line of communication between the adjoining republics of Chili and Argentaria * is over the Uspallata Pass at the head of the valley of Aconcagua ; and Santa Rosa, or as it is more commonly called, Los Andes, is the starting-point for travellers from the west. Don B. V. Mackenna had kindly furnished me with a letter to the officer in charge of the custom-house station at the foot of the pass, known as the *Resguardo del Rio*

---

* The inconvenience of using a periphrasis for the name of so important a country may warrant my adoption of the obvious name Argentaria in place of Argentine territory, or Argentine Confederation, and I shall adhere to the shorter designation in the following pages.

Colorado, and led me to believe that a carriage road extended as far as that point. The latter statement was, however, disputed by several of my acquaintances in Santiago, and the most various assertions were made as to the distance and the time requisite for the excursion. As it turned out, Mr. Mackenna, as he generally is, was correctly informed. The road, as I now learned, was in bad order, but quite passable for a carriage ; and the distance could be accomplished in little over three hours.

Having ordered a vehicle for the next morning, I inquired for a man or a boy acquainted with the neighbourhood of the town, who might serve as guide and carry some of the traps with which a botanist is usually encumbered. An ill-looking fellow, who seemed to have been drinking heavily overnight, soon made his appearance, and we started through a long, dusty street, with only very few houses at wide intervals, which led to the road by which I was to travel on the following morning. Seeing the ground near the town to be much inclosed, while on the opposite side of the river a broad belt of flat stony ground, partly covered with bushes and small trees, gave better prospect to the botanist, I desired to be conducted to the nearest bridge by which I might cross the stream. When we reached the place it appeared to be even a more rickety structure than usual, requiring some care to avoid the numerous holes in the basket-work which formed the floor. Having ascertained that I meant to return the same way, my guide proceeded to stretch himself on the bank, where I found him fast asleep on my return.

The character of the vegetation was the same as that about Santiago, but the general aspect indicated a decided increase of dryness in the climate, so that at the present season there was very little remaining to be gleaned by the botanical collector. As usually happens, however, careful search did not go quite unrewarded. I found several species not before seen, and even where there were no specimens fit for preservation something was to be learned. My next object was to ascend the neighbouring hill, or *cerro*, which immediately overlooks the town of Santa Rosa. A new proprietor had bought a tract of land on the left bank of the river, and erected very substantial fences rather troublesome to a trespasser. My so-called guide dropped behind as I began to ascend the hill—only five or six hundred feet in height—finally turned back, and, having deposited my goods at the hotel, claimed and received an ill-earned fee. The stony slopes were utterly parched, yet I found a few botanical novelties. A small shrubby composite with prickly leaves, but with the habit and inflorescence of a *Baccharis*, was still in tolerable condition. I took it for the female plant of some species of that characteristic South American diœcious genus; but I afterwards ascertained that it belonged to a completely different group, namely, the *Mutisiaceæ*, being the *Proustia baccharoides* of Don.

The view from the summit of the Cerro towards the Andean range was not equal to that from San Felipe, but on the opposite side the outlook towards the plain was interesting. The contrast between the zone of cultivation in the low lands accessible to irrigation and the higher ground, burnt by the summer to a uniform yellow-brown tint, was striking to the eye. The town of Santa Rosa, laid out on the flat at the foot of the hill, was a curious feature in the prospect. It was designed on the regular plan which seems to have recommended itself to all the European settlers in the American continent, but which I have nowhere seen so exactly carried out as at this place. A chess-board supplied the model, with one row of squares cut off to avoid some rough ground. Fifty-six squares—*quadras*—exactly equal in size, are divided by broad roads, and the whole is surrounded by a wall about half a mile in length each way. The *quadra* in the centre forms the *plaza*; the others were to be occupied by houses and gardens. To make the town, as planned by its founders, a perfect model, it wants nothing but houses and people to live in them. It was, perhaps, imagined that, being on the main line of communication across the Andes, this might become a place of some importance; but the traffic is very limited, and, such as it is, it is carried on by trains of horses and mules that travel to and fro between Valparaiso and Mendoza. The area of land fit for cultivation in the valley above San Felipe is small, and the resort of retail traders doubtless very limited. The result is that Santa Rosa is a town without houses. Many of the *quadras* are occupied by a single house and annexed garden, and only round or close to the *plaza* is such a thing as a row of adjoining buildings to be seen.

The morning of May 25 was noteworthy as producing the solitary instance of punctuality in a native of South America that I encountered in the course of my journey. The virtuous driver of the carriage which I had engaged to take me to the Resguardo was actually at the door of the hotel at the appointed hour, soon after sunrise; but it availed little for my object. Not a soul was stirring in the hotel; and though I made no small disturbance, it was long before I could induce the lazy waiter to make his appearance. I had not thought of providing my breakfast overnight, and could not start without food for a long day's expedition.

At length we started on the road by the left bank which I had followed on the previous evening, and, the weather being again nearly perfect, I thoroughly enjoyed a very charming excursion, which carried me farther into the heart of the Cordillera than I had yet reached. As very often happens, however, the nearer one gets to the great peaks the less one is able to see of them. The general outline of the slopes in the inner valleys of high mountain countries is usually convex, because the torrents have deepened the trench between opposite slopes more quickly than subaërial action has worn away the flanks; and it is only exceptionally that the summits of the ridges can be seen from the intervening valley. Among mountains where the main lines of valley are, so to say, *structural* —*i.e.* depending on inequalities produced during the original elevation of the mountain mass—the case is somewhat different. Such valleys are usually nearly straight, as we see so commonly in the European Alps, and the peaks lying about the head of the valley are therefore often in view; but in the Andes, as in many parts of the Rocky Mountains, it would appear that the valleys are exclusively due to erosive action, and, their direction being determined by merely local conditions, they are extremely sinuous, and rarely follow the same direction for any considerable distance.

The road up the Aconcagua valley seemed to me at the time to be about the worst over which I ever travelled in a carriage, but I had not then made acquaintance with the mountain tracks, which they are pleased to call roads, in the United States. Looking back in the light of subsequent experience, I suppose that the Chilian roads should rank among the best in the American continent, although this one was so uneven that in awkward places, where it overhung the river, the carriage was often tilted so much to one side that I was thankful not to have with me a nervous companion.

About half-way to the Resguardo the road crosses the river by a stone bridge, where it rushes in a narrow channel between high rocky banks. Seeing botanical inducements, I descended to examine the banks on either side, and in crossing the bridge noticed, what I might otherwise have overlooked, that the crown of the arch was rapidly giving way. There was a large hole in the centre, and the structure was sustained only by the still solid masonry on each side, where the wear and tear had been less constant. I have often admired the calm good sense displayed by the horses in all parts of America, and was interested in observing the prudent way in which our steeds selected the safest spots on either side of the hole without any appearance of the nervousness which seems hereditary in English horses, partly due, I suppose, to the unnatural conditions in which they live. With every confidence in animal sagacity, but none whatever in the stability of the bridge, I thought it judicious on my return in the evening to recross it on foot.

I found two or three curious plants not before seen on the rocks here, and again found the singular Zygophyllaceous shrub *Porliera hygrometrica*, which is not uncommon in this part of Chili. The numerous stiff spiny branches diverging at right angles must produce flowers during a great part of the year, as I observed at this season both nearly ripe fruit and flowers in various stages of development. The small pinnate leaves, somewhat resembling in form those of the sensitive plant, have something of the same quality. But in this case the effective stimulus seems to be that of light, causing them to expand in sunshine and to close when the sky is covered. If at all, they must be very slightly affected by contact, as I failed to observe it. If I am correct, the appropriate specific name would be *photometrica* rather than *hygrometrica*.

In the hedges and among the bushes a pretty climbing plant (*Eccremocarpus scaber*) seemed to be common on the right bank of the stream, producing flower and ripe fruit at the same time. It belongs to the trumpet-flower tribe (*Bignoniaceæ*), though not rivalling in size or brilliancy of colour the true Bignonias which I afterwards saw in Brazil.

Having passed on the left the opening of a narrow

valley which appears to contain the main stream of the Aconcagua, I reached the Resguardo somewhat before noon, and proceeded at once to deliver my letter to Captain X——, the officer commanding the frontier station. I was most courteously received, with a pressing invitation to join the *almuerzo*, or luncheon, which is the ordinary midday meal in Chili. Besides the lady of the house, I met at table an officer of the Chilian navy, a friend of my host, who had come to recruit in mountain air after recovery from a serious illness, and who spoke English fairly well. The conversation was interesting, and I was struck by the excellent tone and quick intelligence displayed by these agreeable specimens of Chilian society. In the kindest way, and with evident sincerity, my host pressed me to remain for a week at his house, and promised me many excursions in the neighbourhood. It was with real reluctance that, owing to imperious engagements, I was forced to decline the hospitable invitation; and it has been a further regret that, having failed to note it at the time, my treacherous memory has not retained the name of this amiable gentleman.

Meanwhile, although the time passed so pleasantly, I was burning with the desire to make use of the brief interval available for seeing something of the surrounding country. The Resguardo stands at the junction of a rivulet that descends from the Uspallata Pass with the Rio Colorado, which flows from the north-east apparently from the roots of the great peak of Aconcagua. As far as I could see, the track leading to the pass wind in zigzags up steep slopes, at this season almost completely bare of vegetation, and I decided on following the valley of the Rio Colorado, where, at least along the banks of the stream, vegetation was comparatively abundant. My obliging host had provided a horse and a guide, and I rode for about an hour up the valley, which in great part is narrowed nearly to a ravine. In one place, where it widens to a few hundred yards, I passed a peasant's cottage, with a few stony fields from which the crop had been gathered.

Among the plants not before observed, I was at first puzzled by a sort of thicket of long green leafless stems eight or ten feet in height, growing near the stream. Only after searching for some time I detected some withered remains of a short spike of flowers at the ends of the stems, which showed the plant to be of the *Verbena* family. Whatever may be the original home of that ancient tribe which has spread throughout all the temperate and tropical regions of the earth, it is in South America, and especially in the extra-tropical regions, that it has developed the greatest variety both of genera and species. On the heights of the Peruvian Andes, from the snows of the Chilian Cordillera to the shores of the Pacific, as well as on the plains of Argentaria and Uruguay, the botanist is everywhere charmed by the brilliant flowers of numerous species of true *Verbena*. In the warmer zone the allied genus *Lippia* becomes predominant, and displays an equal variety of aspect; but in Chili especially we find a number of plants very different in aspect, although nearly allied in structure to the familiar types. The plant of the Rio Colorado—

known to botanists as *Baillonia spartioides*—appears to be rare in Chili, as it is not among the species collected by the earlier explorers of this region.

I was interested in finding here two species of *Loranthus*, which, unlike their congeners, grow in a respectable way, depending on their own resources for subsistence. The great majority of nearly four hundred known species of this genus live as parasites on the stems of other plants, but these form bushes with woody roots, which apparently have not even an underground connection with those of their neighbours. When I returned to the Resguardo, laden with plants, it was high time to think of starting homeward to Santa Rosa. I did not much fancy travelling by night over the curious road that I had followed in the morning, and my coachman seemed to hold the same opinion very strongly. Accordingly I soon started, after cordial leave-taking, but was a little surprised when, without previous warning, the driver pulled up his horses at the garden gate of a substantial house, which I had noticed in the morning a few hundred yards below the Resguardo. Presently a young man came out, and, addressing me in very fair English, explained that he had written to order a carriage for the following day, but would be thankful if I could give him a seat to convey him to his family at Santa Rosa. Of course I willingly consented, and in the conversation, which was carried on alternately in Spanish and English during the following three hours, I gained an opportunity for some practice in a language which has never been quite familiar to me.

I became interested in the poor young fellow, who

was evidently in an advanced stage of pulmonary consumption. He had been on a visit with friends, in the vain hope that the pure air of this mountain valley might arrest the disease, and now, as the season was far advanced, wished to rejoin his wife and children at Santa Rosa. Like many consumptive patients, he had a feverish proneness for talk ; and, having first told me his own story, he asked me a multitude of questions respecting my present journey and as to the other countries that I have visited. At length, with evident reference to my age, he gravely said, "No le parece Señor que es tiempo para descansar?" I answered that there would be time enough to *descansar* when one is laid underground, and that for the present I saw no occasion to rest. As I stopped the carriage only two or three times to gather plants, and the driver kept his horses at a smart trot most of the way, we accomplished the return journey of eighteen or twenty miles in a little under three hours, and reached the town at nightfall.

On the 26th I returned to Valparaiso, meeting the Santiago train at the now familiar junction station of Llaillai. Although the weather was still fine, clouds hung round the Cordillera, and I was not destined again to enjoy the glorious view of the great range. My first care on reaching the port was to secure my passage in the German steamer as far as the Straits of Magellan. I found that the steamship *Rhamses* of the Cosmos line, which in ordinary course should have departed on the 28th, was delayed until the following day, May 29. It was inevitable to regret that the additional day had not been devoted to the Rio

Colorado, but, in fact, I found my time fully occupied during the two days that remained available. The collections of dried plants made up to this time had to be packed securely in the chest in which they were to remain until they reached England, and, as every botanist knows, it is expedient to hasten the process of drying fresh plants as far as possible before going to sea, where the operation is always one of difficulty.

I was invited to dinner on the day of my arrival by Mr. C——, one of the chief English merchants established in Chili, and acquired some interesting information from his conversation. Having been at work during a great part of the previous night, I was, however, thoroughly tired, and was able to profit less than I should have done by the hospitable entertainment. On the morning of my departure from Cauquenes I had met Mr. Edwin Reed, an English naturalist many years resident in Chili, and by appointment called upon him at his house in Valparaiso. Mr. Reed has a good knowledge of the botany and zoology of his adopted country, and several hours were agreeably spent on each of the two available days in going through parcels of his duplicate collections, when he was good enough to give me flowering specimens of plants which I had seen only in imperfect condition, as well as of many others from the higher region of the Cordillera which had been entirely inaccessible to me.

My visit to Chili had now come to an end. All needful preparations were concluded; and, after a busy morning and an excellent luncheon at the Hotel Colon, I went on board the *Rhamses* early in the afternoon of May 29, not without deep regret at quitting a country where I had spent twenty of the most enjoyable days of my life. The only occupants of the first-class saloon were a German gentleman, Mr. Z———; his wife, a delicate Peruvian lady, who remained in her cabin during most of the voyage; five children; and a maid. I found a good clean cabin, which had been reserved for my use, and before long a tall, handsome man of pleasant countenance introduced himself to me as Captain Willsen, commanding the *Rhamses*.

The steamers of the German Cosmos line, of which this is, I believe, a fair example, differ in many respects from the great English ocean steamships which conduct most of the intercourse between Europe and South America. They are mainly destined for cargo, the accommodation for passengers being comparatively very limited, and of scarcely half the dimensions, being of rather less than two thousand tons displacement by our measurement. In our passenger ships speed is always the foremost consideration. In accordance with the national temperament, the German steamers set slight store on that object; safety and economy are the aims constantly kept in view, and the consumption of an increased quantity of coal in order to gain a day would be regarded as culpable extravagance. The especial advantage which they offer to every traveller in this region is that, owing to their light draught, they are able to traverse the narrow and intricate channels of Western Patagonia between the mountainous islands and the mainland; while to sea-sick passengers the object of avoiding more than four hundred miles of the heavy seas of the Southern Pacific is a further inducement. A naturalist finds an additional attraction in the general sympathy and helpfulness which he may expect from every officer in a German ship. Courtesy and friendly feeling are almost invariably to be found on board our steamers, but the pursuits of a naturalist rarely seem to call forth the slightest show of interest.

Our departure was fixed for two p.m., but in fact we did not move till past seven, long after dark at this season. On getting out to sea we found a moderate swell running from the southward, and moved slowly, as coal was economized. On the following morning we found ourselves rather far from land, and, although the weather was moderately clear, we had only a few distant glimpses of the coast during the day. The barometer fell slowly about two-tenths of an inch from morning to night, and it seemed evident that we were about to bid farewell to the bright skies of Central Chili. We were to take in coal for the voyage to Europe at Lota, about two hundred and fifty nautical miles south of Valparaiso. That distance could be easily accomplished, even by the *Rhamses*, in twenty-four hours; but as there was no object in arriving before morning, we economized fuel and travelled slowly. Heavy rain fell during the entire night, and ceased only when, on the morning of May 31, we entered the harbour of Lota.

Lota is a place which, although not marked on Stanford's latest map of South America, has within a short time risen to considerable importance, owing to the discovery of extensive deposits of lignite of excellent quality. I have heard various estimates of its value as steam coal, the lowest of which set five tons of Lota coal as equal to four of Welsh anthracite. The seams appear to be of considerable thickness, and the underground works have now extended to a considerable distance from the shore. All the ocean steamers returning to Europe now call here for their provision of fuel, and in addition the proprietor has established extensive works for smelting copper and for making glass. The owner of this great property is a lady, the widow of the late Mr. Cousiño, whose income is rated at about £200,000 a year. About 2500 people are constantly employed, who, with their families, inhabit a small town of poor appearance which has grown up on the hill overlooking the harbour.

I was courteously invited to the house of Mr. Squella, a relation of Madame Cousiño, who has the direction of this great establishment, and there had the pleasure of again meeting my former travelling companion, Mr. H——, and also Captain Simpson, an officer of the Chilian navy of English extraction, who, while commanding a ship on the southern coast, has rendered some services to science. The conversation was carried on chiefly in English, which has decidedly become the *lingua franca* of South America, but was shortened by my natural anxiety to turn to the best account the short time at my disposal. I had a choice between three alternatives—a descent into the coal mine, a visit to the works above ground and the miners' town, or a ramble through the so-called park, which occupies the promontory stretching westward which forms the natural harbour of Lota, and covers a great portion of the precious deposit to which the place owes its new-born importance. I naturally preferred the latter, feeling that my limited experience as a geological observer would not allow me to profit much by a subterranean excursion. I made inquiry, however, as to the vegetable remains found in the lignite, and I was told that they are abundant, although the few specimens which I saw showed but slight traces of vegetable structure. I was led to believe that a collection of specimens had been sent to Europe to my late lamented friend, Dr. Oswald Heer, but I am not aware that he has left any reference to such a collection, or even that it ever reached his hands.

The *parque* of Lota, to which I directed my steps, has rather the character of an extensive pleasure-ground than of what we call a park; but the surface is so uneven, and the outline so irregular, that I could not estimate its extent. The numerous fantastic structures in questionable taste that met the eye in every direction create at the first moment an unfavourable impression, but the charms of the spot are so real that this is soon forgotten. The variety and luxuriance of the vegetation, and the diversified views of the sea and the rocky shores, were set off by occasional bursts of bright sunshine, in which the drops that still hung on every leaflet glittered like jewels of every hue. The trees here were of very moderate dimensions, the largest (here called *roble*) being of the laurel family, which, for want of flower or fruit, I failed to identify. The Spaniards in South

America have given the name *roble*, which properly means "oak," to a variety of trees which agree only in having a thick trunk and spreading branches. The shrubs were very numerous, partly indigenous and partly exotic, and a peculiar feature which I have not noticed in any other large garden is the number of parasites living on the trunks and branches of the trees and shrubs. Ferns were very numerous and grow luxuriantly, showing a wide difference of climate between this coast and that of the country two or three degrees further north. But the great ornament of this place is the beautiful climber, *Lapageria rosea*, now producing in abundance its splendid flowers, which so finely contrast with its dark-green glossy foliage. The specific name *rosea* is unfortunate, as the colour of the flowers is bright crimson, verging on scarlet.

One of the special features of this garden was the abundance of humming-birds that haunted the shrubs and small trees, and darted from spray to spray with movements so rapid that to my imperfect vision their forms were quite indistinguishable. Whenever I drew near in the hope of gaining a clearer view, they would dart away to another shrub a few yards distant, and I am unable to say whether the bright little creatures belonged to one and the same or to several different species.

At one place where the garden is only some twenty feet above the beach, I scrambled down the rocks, and was rewarded by the sight of two or three plants characteristic of this region. The most attractive of these is one of the many generic types peculiar to

P

the Chilian flora, allied to the pine-apple. The long stiff leaves, edged with sharp teeth and radiating from the lower part of the stem, are coloured bright red along the centre and at the base, forming, when seen from a distance, a brilliant, many-rayed red star. Another novelty was *Francoa sonchifolia*, which also clings to the rocks by the sea. It has somewhat the habit of a large crucifer, but the structure of the flower and fruit is widely different. It was regarded by Lindley as the type of a distinct natural family, but has been, with one other Chilian genus (*Tetilla*), classed as a tribe of the saxifrage family.

Time passed quickly in such an interesting spot, and the hour appointed for returning to the ship had nearly arrived, when Mr. Reilly, the gardener who has the management of the *parque*, invited me to see his house. He came, as I learned, from Wexford, in Ireland, had had some training in the Royal Gardens at Kew, when his fortunate star led him to Chili. I found him installed in a very pretty and comfortable house, charmingly situated, in as full enjoyment of one of the most beautiful gardens in the world as if he were its absolute owner. This was only one more instance of the success which so often attends my countrymen when removed to a distance from their native land. Freed from the evil influences that seem indigenous to the soil of that unfortunate island, they develop qualities that are too rarely perceptible at home. The arguments for emigration are commonly based only on the economical necessity for relieving the land of surplus population; to my mind it may be advocated on other and quite different grounds. For every Irishman who is carried to a distant land there is a strong probability of a distinct gain to the world at large.

I left the *parque* at Lota with my memory full of pictures of a spot which, along with Mr. Cooke's famous garden at Montserrat, near Cintra, and that of M. Landon in the oasis of Biskra, I count as the most beautiful garden that I have yet seen.

A rather large island—Isla de Sta. Maria—lies off the Chilian coast to the west of Lota, and is separated on the southern side from the promontory of Lavapie by a channel several miles wide. But as this is beset with rocks, the rule of the German steamers is to avoid the passage, excepting in clear weather by day. In deference, therefore, to this cautious regulation, we set our helm to the north on leaving Lota, two or three hours after sunset, and only after keeping that course for some ten miles, and running past the small port of Coronel, steered out to seaward, and finally resumed our proper southerly direction. Our sleep was somewhat disturbed by the heavy rolling of the ship during the night, and the morning of the 1st of June broke dimly amid heavy lowering clouds, just such a day as one might expect at the corresponding date (December 1) on the western coast of Europe. Although the sea was running high, there was little wind. The barometer at daybreak stood at 29·98, having risen a tenth of an inch since the previous evening, and the temperature was about 52° Fahr. In our seas one would suppose that a gale must have recently prevailed at no great distance, but I believe the fact to be that in the Southern Pacific high seas prevail during a great part of the year, even where no strong winds are present to excite them. Gales are undoubtedly common in the zone between the fiftieth and sixtieth degrees of south latitude, and the waves habitually run higher there than they ever do in the comparatively confined area of the Atlantic. The disturbances are propagated to great distances, modified, of course, by winds, currents, and the form of the coasts when they approach the land; but the smooth waters that extend more than thirty degrees on either side of the equator are rarely encountered in higher latitudes. The skies brightened as the day wore on, and the sun from time to time broke through the clouds; but we were out of sight of land, and the only objects in view during the day were the sea, the sky, and the numerous sea-fowl that followed the ship. The incessant rolling made it difficult to settle down to any occupation.

We were now abreast of that large tract of Chili which has been left in the possession of its aboriginal owners, the Araucanian Indians, extending about one hundred miles from north to south, and a rather greater distance from the coast to the crest of the Cordillera. It is unfortunate that so little is known of the Araucanians, as, in many respects, they appear to be the most interesting remaining tribe of the aboriginal American population. For nearly two centuries they maintained their independence in frequent sanguinary encounters with the Spaniards, which are said on Chilian authority to have cost the invaders the loss of 100,000 men. Since the establishment of Chilian independence, the policy of the republic has been to establish friendly relations with this indomitable people. The territory between the Bio-Bio river to the north and the Tolten to the south was assigned to them, and small annual donations were made to the principal chiefs on condition of their maintaining order amongst the tribesmen. During the last forty years, however, white settlers have trespassed to a considerable extent on the Indian territory, both on the north and south sides, but have generally contrived to keep up friendly intercourse with the natives, while Chilian officials, established at Angol on the river Mallego, exercise a species of supervision over the entire region.

The present Araucanian population is somewhat vaguely estimated at about 40,000, and it is a question of some interest whether, like most native races in contact with those of European descent, they will ultimately be improved out of existence, or be gradually brought within the pale of civilization and fused with the intrusive element. The soil is said to be in great part fertile; they raise a large quantity of live stock, and some of the chiefs are said to have amassed wealth, and to have begun to show a taste for the comforts and conveniences of civilized life.

While at Santiago, I made some inquiry as to the language of the Araucanian tribes. I was informed that in the seventeenth century the Jesuit missionaries published a grammar of the language, of which only two or three copies are known to exist. About the beginning of this century a new edition, or reprint, of this work appeared at Madrid, but, as I was assured, has also become extremely rare, and copies are very seldom to be procured.

On the evening of the 1st the barometer had risen about a tenth of an inch, but by the following morning had returned to the same point (very nearly thirty inches) as on the previous day, without any change in the state of the weather; but we enjoyed more sunshine, and the proceedings of the birds that ceaselessly bore us company afforded us constant occupation and amusement. Two species were predominant. One of these was the well-known cape pigeon (*Daption capensis*), familiar to all mariners in the southern hemisphere. This is a handsome bird, much larger than a pigeon, exhibiting a considerable variety of plumage in what appeared to be adult individuals. In all the ground colour is white, and the tips of the spreading tail feathers are dark brown or nearly black. The upper surface of the wings sometimes showed a somewhat tesselated pattern of white and dark brown, but more commonly were marked by two transverse dark bands, with pure white between. They were very numerous, as many as from fifty to a hundred being near the ship at the same time, keeping close company, and often swooping over the deck a few feet over our heads; but, although seemingly fearless, they never were induced to take a piece of meat from a man's hand, though the temptation was often renewed. The next in frequency—called on this coast *colomba*—is nearly as large as the cape pigeon, with plumage much resembling that of a turtle dove. This also approached very near. Both of these birds seemed to feel fatigue, as, after circling round the ship for half an hour at a time, they would rest on the surface of the water, dropping rapidly astern, but after some minutes resume their flight and soon overtake the ship. More interesting to me were the two species of albatross, which I had never before had an opportunity of observing. These were more shy in their behaviour, never, I think, approaching nearer than seventy or eighty yards, and usually following the ship with a slow, leisurely flight still farther astern. The common, nearly white, species (*Diomedea exulans*) is but a little larger than the dark-coloured, nearly black species, which I supposed to be the *Diomedea fuliginosa* of ornithologists.* If, as is probable, the same birds followed us all day, we saw but two of the latter, which are, I believe, everywhere comparatively scarce. In both species I was struck by the peculiar form of the expanded wing, which is very narrow in proportion to its great length.

The moment of excitement for the birds, as well as for the lookers-on, was when a basket of kitchen refuse was from time to time thrown overboard. It was amusing to watch the rush of hungry creatures all swooping down nearly at the same point, and making a marvellous clatter as they eagerly contended for the choice morsels. It did not appear to me that the smaller birds showed any fear of the powerful albatross, or that the latter used his strength to snatch away anything that had been secured by a weaker rival.

About noon on the 2nd of June we were abreast of the northern part of the large island of Chiloe, but were too far out to sea to get a glimpse of the high

---

* It is quite possible that the bird which I took for the black albatross was the giant petrel, common, according to Darwin, in these waters, and closely resembling an albatross.

land on the west coast. At the northern end the island is separated from the mainland by a narrow channel (Canal de Chacao) only two or three miles in width ; but on the east side the broad strait or interior sea between Chiloe and the opposite coast is from thirty to forty miles in breadth, and beset by rocky islets varying in size from several miles to a few yards.

Another unquiet night ushered in the morning of the 3rd of June. This was fairly clear, with a fresh breeze from the south-west, which, as the day advanced, rose nearly to a gale. The sea did not appear to run higher than before, but the waves struck the ship's side with greater force, and at intervals of about ten minutes we shipped rather heavy seas, after which the deck was nearly knee-deep in water, and a weather board was needed to keep the saloon from being flooded. The barometer fell slightly, and the temperature was decidedly lower, the thermometer marking about 50° Fahr. Some attempts at taking exercise on the hurricane deck were not very successful, my friend, Mr. H——, being knocked down and somewhat bruised, and we finally retired to the saloon, and found the state of things not exhilarating. We saw nothing of the Chonos Archipelago, consisting of three large and numerous small islands, all covered with dense forest, and separated from the mainland by a strait, yet scarcely surveyed, about a hundred and twenty miles in length, and ten to fifteen in breadth.

Darwin, writing nearly fifty years ago, anticipated that these islands would before long be inhabited, but I was assured that no permanent settlement has ever been established. Parties of woodcutters have from time to time visited the islands, but no one has been tempted to remain. The excessive rainfall, which is more continuous in summer than in winter, makes them unfit for the residence of civilized man; but it seems probable that Fuegians transported there would find conditions favourable to their constitution and habits of life. It is another question whether the world would be any the better for the multiplication of so low a type of humanity.

In the afternoon, as the sea was running very high, the captain set the ship's head to the wind. We saw him but once, and perceived an anxious expression on his usually jovial countenance. It afterwards came out that he apprehended the continuance of the gale, in which case he might not have ventured to put the helm round so as to enter the Gulf of Peñas. At nightfall, however, the wind fell off, and by midnight the weather was nearly calm, though the ship gave us little rest from the ceaseless rolling. During all this time sounds that issued at intervals from the cabin of the Peruvian lady and her children showed that what was merely a bore to us was to them real misery. I have often asked myself whether there is something about a sea-voyage that develops our natural selfishness, or whether it is because one knows that the suffering is temporary and has no bad results, that one takes so little heed of the really grievous condition of travellers who are unable to bear the movement of the sea. A voyage with sea-sick passengers, especially in bad weather, when one is confined to the saloon, is a good deal like being lodged in one of the prisons of the Spanish inquisition while torture was freely applied to the unhappy victims; and yet persons who are not counted as hard-hearted seem to bear their position with perfect equanimity, if not with something of self-satisfaction.

The morning of the 4th of June was so dark that we supposed our watches to have gone astray. Of course, the days were rapidly growing shorter as we ran to the southward, but the dim light on this morning was explained when we sallied forth. The wind had veered round to the north, and in these latitudes that means a murky sky with leaden clouds above and damp foggy air below. The change, however, was opportune. We were steering about due southeast, entering the Gulf of Peñas, with the dim outline of Cape Tres Montes faintly seen on our larboard bow.

I have already alluded to the peculiar conformation of the south-western extremity of the South American continent, which, from the latitude of 40° south to the opening of the Straits of Magellan, a distance of about nine hundred miles, exhibits an almost continuous range of high land running parallel to the southern extremity of the great range of the Andes. At its northern end this western range, under the names Cordillera Pelada and Cordillera de la Costa, forms part of the mainland of Chili, being separated from the Andes by a broad belt of low country including several large lakes, those of Ranco and Llanquihue being each about a hundred miles in circuit. South of the Canal de Chacao the range is continued by the island of Chiloe and the Chonos Archipelago, and then by the great mountainous promontory whose southern extremity is Cape Tres Montes. Here occurs the widest breach in the continuity of the range, as the Gulf of Peñas is fully forty miles wide. To the southward commences the long range of mountainous islands that extend to the Straits of Magellan, between which and the mainland lie the famous channels of Western Patagonia. It is worthy of note that, corresponding to the elevation of this parallel western range, the height of the main chain of the Andes is notably diminished. Of the summits that have hitherto been measured south of latitude $42°$ only one—the Volcano de Chana—attains to a height of eight thousand feet, and there is reason to believe that numerous passes of little more than half that elevation connect the eastern and western slopes of the chain.

Another point of some interest is the northern extension of the so-called antarctic flora throughout the whole of the western range, many of the characteristic species being found on the Cordillera Pelada close to Valdivia, which does not, I believe, much exceed three thousand feet in height. It is true that a few antarctic species have been found in the higher region of the Andes as far north as the equator, just as a few northern forms have travelled southward by way of the Rocky Mountains and the highlands of Mexico and Central America; and Professor Fr. Philippi has lately shown that many southern forms, and even a few true antarctic types, extend to the hills of the desert region of Northern Chili, where the constant presence of fog supplies the necessary moisture.*

* See an interesting paper in the *Journal of Botany* for July, 1884.

The true northern limit, however, of the antarctic flora may be fixed at the Cordillera Pelada of Valdivia.

We crept on cautiously into the gulf, anxiously looking out for some safe landmark to secure an entrance into the northern end of Messier's Channel. Soon after midday we descried a remarkable conical hill, which is happily placed so as to distinguish the true opening from the indentations of the rocky coast. As we advanced the air became thicker and colder, as drizzling rain set in; but the practised eyes of seamen are content with indications that convey no meaning to an ordinary landsman, and just as the night was closing in almost pitch dark, the rattle of the chain cable announced that we had come to anchor for the night in Hale Cove.

The weather had become very cold. At two p.m. in the gulf the thermometer stood at 42°, and after nightfall it marked only a few degrees above freezing-point, so that, even in the saloon, we sat in our great coats, not at all enjoying the unaccustomed chilliness. All rejoiced, therefore, when the captain, having quite recovered his wonted cheerfulness, announced that a stove was to be set up forthwith in the saloon, and a tent erected on deck to give shelter from the weather. The stove was a small, somewhat rickety concern, and we fully understood that it would not have been safe to light it while the ship was labouring in the heavy seas outside; but it was especially welcome to me, as I was anxiously longing for the chance of getting my botanical paper thoroughly dry. As we enjoyed a cheerful dinner, two of the officers pushed off in one of the ship's boats into the blackness that had closed around. After some time a large fire was seen blazing a few hundred yards from the ship, and, amid rain and sleet, we could descry from the deck some moving forms. They had succeeded, I know not how, in getting the damp timber into a blaze, and were good-naturedly employed in gathering whatever they could lay hands upon to contribute to my botanical collection. Not much could be expected under such conditions, but everything in this, to me, quite new region was full of interest. Dead branches covered with large lichens introduced me to one of the most characteristic features of the vegetation. The white fronds, four or five inches wide, and several feet in length, enliven the winter aspect of these shores, and possibly supply food to some of the wild animals. Among the plants which had been dragged up at random were several roots of the wild celery of the southern hemisphere. It is widely spread throughout the islands of the southern ocean, as well as on the shores of both coasts of Patagonia, and was described as a distinct species by Dupetit Thouars; but in truth, as Sir Joseph Hooker long ago remarked in the "Flora Antarctica," there are no structural characters by which to distinguish it from the common wild celery of Europe, which is likewise essentially a maritime plant. Growing in a region where it is little exposed to sunshine, it has less of the strong characteristic smell of our wild plants, and the leaves may be eaten raw as salad, or boiled, which is not the case with our plant until the gardener, by heaping earth about the roots, diminishes the pungency of the smell and flavour.

One thought alone troubled me as I lay down in

my berth to enjoy the first quiet night's rest. If the weather should hold on as it now fared, there was but a slight prospect of enjoying the renowned scenery of the channels, or of making much acquaintance with the singular vegetation of this new region. It was therefore with intense relief and positive delight that I found, on sallying forth before sunrise, a clear sky and a moderate breeze from the south. Snow had fallen during the night, and was now hard frozen; and in the tent, where my plants had lain during the night, it was necessary to break off fragments of ice with numbed fingers before laying them in paper.

We weighed anchor about daybreak, and the 5th of June, my first day in the Channels, will ever remain as a bright spot in my memory. Wellington Island, which lay on our right, is over a hundred and fifty miles in length, a rough mountain range averaging apparently about three thousand feet in height, with a moderately uniform coast-line. On the other hand, the mainland presents a constantly varying outline, indented by numberless coves and several deep narrow sounds running far into the recesses of the Cordillera. In the intermediate channel crowds of islets, some rising to the size of mountains, some mere rocks peeping above the water, present an endless variety of form and outline. But what gives to the scenery a unique character is the wealth of vegetation that adorns this seemingly inclement region. From the water's edge to a height which I estimated at fourteen hundred feet, the rugged slopes were covered with an unbroken mantle of evergreen trees and shrubs. Above that height the bare declivities were clothed with snow, mottled at first by projecting rocks, but evidently lying deep upon the higher ridges. I can find no language to give any impression of the marvellous variety of the scenes that followed in quick succession against the bright blue background of a cloudless sky, and lit up by a northern sun that illumined each new prospect as we advanced. At times one might have fancied one's self on a great river in the interior of a continent, while a few minutes later, in the openings between the islands, the eye could range over miles of water to the mysterious recesses of the yet unexplored Cordillera of Patagonia, with occasional glimpses of snowy peaks at least twice the height of the summits near at hand. About two o'clock we reached the so-named English Narrows, where the only known navigable channel is scarcely a hundred yards in width between two islets bristling with rocks. The tide rushed through at the rate of a rapid river, and our captain displayed even more than his usual caution. Some ten men of the crew were posted astern with steering gear, in readiness to provide for the possible breakage of the chains from the steering-house. It seemed unlikely enough that such an accident should occur at that particular point, but there was no doubt that if it did a few seconds might send the ship upon the rocks.

One of the advantages of a voyage through the Channels is that at all seasons the ship comes to anchor every night, and the traveller is not exposed to the mortification of passing the most beautiful scenes when he is unable to see them. When more thoroughly known, it is likely that among the numerous coves many more will be found to offer good anchorage; but few are now known, and the distance that can be run during the short winter days is not great. We were told that our halt for the night was to be at Eden Harbour, less than twenty miles south of the English Narrows, and to my great satisfaction we dropped anchor about 3.30 p.m., when there was still a full hour of daylight. Our good-natured captain put off dinner for an hour, and with all convenient speed I went ashore with Mr. H—— and two officers of the ship.

Eden Harbour deserves its name. A perfectly sheltered cove, with excellent holding-ground, is enclosed by steep forest-clad slopes, culminating to the north in a lofty conical hill easily recognized by seamen. The narrow fringe between the forest and the beach is covered with a luxuriant growth of ferns and shrubby plants, many of them covered in summer with brilliant flowers, blooming in a solitude rarely broken by the passage of man. After scrambling over the rocks on the beach, the first thing that struck us was the curious nature of the ground under our feet. The surface was crisp and tolerably hard, but each step caused an undulation that made one feel as if walking on a thick carpet laid over a mass of sponge. Striking a blow with the pointed end of my ice-axe, it at once pierced through the frozen crust, and sank to the hilt over four feet into the semifluid mass beneath, formed of half-decomposed remains of vegetation.

At every step plants of this region, never before seen, filled me with increasing excitement. Several were found with very tolerable fruit, and there were even some remains of the flowers of *Desfontainea spinosa* and *Mitraria coccinea*. The latter beautiful shrub appears to have been hitherto known only from Chiloe and the Chonos Archipelago. In those islands it is described as a tall climber straggling among the branches of trees. Here I found it somewhat stunted, growing four or five feet high, with the habit of a small fuchsia. Neither of these is a true antarctic species. Like many Chilian plants, they are peculiar and much-modified members of tribes whose chief home is in tropical America. Everything else that I saw was characteristically antarctic. Three small coniferous trees peculiar to this region; a large-flowered berberry, with leaves like those of a holly, growing six or eight feet high, still showing remains of the flower; and two species of *Pernettya*, with berries like those of a bilberry, and which replace our *Vaccinia* in the southern hemisphere, were among the new forms that greeted me.

A few minutes' stumbling over fallen timber brought us to the edge of the forest, and it was soon seen that, even if time allowed, it would be no easy matter to penetrate into it. The chief and only large tree was the evergreen beech (*Fagus betuloides* of botanists). This has a thick trunk, commonly three or four feet in diameter, but nowhere, I believe, attains any great height. Forty feet appeared to me the outside limit attained by any that I saw here or elsewhere. But perhaps the most striking, and to me unexpected, feature in the vegetation was the abundance and luxuriance of the ferns that inhabit these coasts. From out of the stiff frozen crust under our feet a profusion of delicate filmy ferns (*Hymenophylla*) grew to an unaccustomed size, including several quite distinct species; while here and there clumps of the stiff fronds of *Lomaria magellanica*, a couple of feet in height, showed an extraordinary contrast in form and habit. As Sir Joseph Hooker long ago remarked, the regular rigid crown of fronds issuing from a thick rhizome, when seen from a little distance, remind one forcibly of a *Zamia*. It was to me even more surprising to find here in great abundance a representative of a genus of ferns especially characteristic of the tropical zone. The *Gleichenia* of these coasts differs sufficiently to deserve a separate specific name, but in general appearance is strikingly like that which I afterwards saw growing in equal abundance in Brazil.

This continent, with its thousands of miles of unbroken coast-line, and its mountain backbone stretching from the equator to Fuegia, has offered extraordinary facilities for the diffusion of varied types of vegetation. As I have already remarked, some species of antarctic origin travel northward, and some others, now confined to the equatorial Andes, are most probably modified descendants from the same parent stock; while a small number of tropical types, after undergoing more or less modification, have found their way to the extreme southern extremity of the continent.

By a vigorous use of my ice-axe, which is an excellent weapon for a botanist, I succeeded in uprooting a good many plants from the icy crust in which they grew; but the minutes slipped quickly by, daylight was fading in this sheltered spot, shut out from the north and west by steep hills, and too soon came the call to return to the ship. On the beach I picked up the carapace of a crab—bright red and beset with sharp protuberances—evidently freshly feasted on by some rapacious animal. The whole of the body and the shell of the under part as well as the claws had disappeared, leaving nothing but the carapace, which I presume had been found too hard and indigestible. Darwin informs us that the sea-otter of this region feeds largely on this or some allied species of crab.

The cold was sufficient to make the little stove in the saloon of the steamer very acceptable, but at no time throughout the voyage could be called severe. Between noon and three p.m. on the 5th of June the thermometer in the open air stood about 40° Fahr., and fell at night only two or three degrees below freezing-point. The barometer was high, gradually rising from 30 inches to 30.3, at which it stood on the following day. Everything promised settled weather, and it was therefore disappointing to find the sky completely covered when I went on deck early in the morning of the 6th. A light breeze from the north raised the temperature by a few degrees and brought the clouds. The scenery throughout the day was even of a grander character than before, and the absence of sunshine gave it a sterner aspect. At times, when passing the smaller islands, I was forcibly reminded of the upper lake of Killarney, the resemblance being much increased by the appearance of the smaller islets and rocks worn down and rounded by floating ice. On this and the following days I frequently looked out for evidences of ice-action on the rocky flanks of the mountains. These were at some points very perceptible up to a considerable height; but all that I could clearly make out appeared to be directed from south to north, and nearly or quite horizontal. I failed to trace any indication on the present surface of the descent in a westerly direction of great glaciers flowing from the interior towards the coast.

Before midday we passed opposite the opening of Eyre Sound, one of the most considerable of the numerous inlets that penetrate the mountains on the side of the mainland. This is said to extend for forty or fifty miles into the heart of the Cordillera, and it seems certain that one, or perhaps several, glaciers descend into the sound, as at all seasons masses of floating ice are drifted into the main channel. We did not see them at first, as the northerly breeze had carried them towards the southern side of the inlet; but before long we found ourselves in the thick of them, and for about a mile steamed slowly amongst floating masses of tolerably uniform dimensions, four or five feet in height out of the water, and from ten to fifteen feet in length. At a little distance they looked somewhat like a herd of animals grazing. Seen near at hand, the ice looked much weathered, and it may be inferred that the parent glacier reaches the sea somewhere near the head of the sound, and they had been exposed for a considerable time before reaching its mouth.

The existence of great glaciers descending to the sea-level on the west coast of South America, one of which lies so far north as the Gulf of Peñas, about 47° south latitude, is a necessary consequence of the rapid depression of the line of perpetual snow on the flanks of the Andes, as we follow the chain southward from Central Chili to the channels of Patagonia. The circumstance that permanent snow is not found lower than about fourteen thousand feet above the sea in latitude 34°, while only 8° farther south the limit is about six thousand feet above the sea-level, has been regarded as evidence of a great difference of climate between the northern and southern hemispheres, and more especially of exceptional conditions of temperature affecting this coast. It appears to me that all the facts are fully explained by the extraordinary increase of precipitation from the atmosphere, in the form of rain or snow, which occurs within the zone where the rapid depression of the snow-line is observed. So far as mean annual temperature of the coast is concerned, the diminution of heat in receding from the equator is less than the normal amount, being not quite 5° Fahr. for 7° of latitude between Valparaiso and Valdivia. But the annual rainfall at Valdivia is eight times, and at Ancud in Chiloe more than nine times, the amount that falls at Santiago. Allowing that the disproportion may be less great between the snowfall on the Cordillera in the respective latitudes of these places, we cannot estimate the increased fall about latitude 40° at less than four times the amount falling in Central Chili. When we further recollect that in the latter region the sky is generally clear in summer, and that the surface is exposed to the direct rays of a sun not far from vertical, while on the southern coast the sun is constantly veiled by heavy clouds, it is obvious that all the conditions are present that must depress the snow-line to an exceptional extent, and allow of those accumulations of snow that give birth to glaciers. When a comparison is drawn between South Chili and Norway, it must not be forgotten that at Bergen, where the Norwegian rainfall is said to be at its maximum, the annual amount is sixty-seven inches, or exactly one-half of that registered in Chiloe.

It is a confirmation of this view of the subject that in going southward from the parallel of 42° to Cape Froward in the Straits of Magellan, through 12° of latitude, while the fall of mean yearly temperature must be reckoned at 8° Fahr., the depression of the snow-line cannot exceed three thousand feet.* Of course, we have no direct observations of rainfall in the Channels or on the west side of the Straits of Magellan, but there is no doubt that it diminishes considerably in going southward.

To the south of Eyre Sound the main channel opens to a width of four or five miles, and is little encumbered by rocky islets, so that we kept a direct course a little west of south, and in less than two hours reached the southern extremity of Wellington Island, and gained a view of the open sea through a

---

* The estimates given by Pissis do not rest on accurate observations, and seem to me exaggerated. I should be inclined to reckon the difference of height of the snow-line between the extreme stations as nearer to two thousand than to three thousand feet.

broad strait which is known as the Gulf of Trinidad. Now that this has been well surveyed, it offers an opportunity for steamers bound southward that have missed the entrance to the Gulf of Peñas to enter from the Pacific, and take the course to the Straits of Magellan through the southern channels.

We had now accomplished the first stage in the voyage through the Channels. Many local names have been given to the various passages open to navigation on this singular coast; but, speaking broadly, the northern portion, between Wellington Island and the mainland, is called Messier's Channel; the middle part, including a number of distinct openings between various islands, is known as the Sarmiento Channel; and the southern division, between Queen Adelaide Island and the continent, is Smyth's Channel. Facing the Pacific to the south of Wellington Island are three of large size—Prince Henry Island, Madre de Dios, and Hanover Island, besides countless islets which beset the straits that divide these from each other; and the course followed by the steamers lies between the outer islands and another large one (Chatham Island) which here rose between us and the mainland.

In the afternoon the north wind freshened; as a result, the weather became very thick, and rain set in, which lasted throughout the night. Our intended quarters were in a cove called Tom Bay; but our cautious captain, with a due dislike to "dirty weather," resolved to halt in a sheltered spot a few miles farther north, known as Henderson's Inlet. Both these places afford excellent shelter, but the bottom is rocky, and ships are much exposed to lose their anchors. Although we arrived some time before sunset, the evening was so dark, and the general aspect of things so discouraging, that no one suggested an attempt to go ashore. Although we were quite near to land, I could make out very little of the outlines; and, indeed, of this middle portion of the voyage I have retained no distinct pictures in my memory.

It struck me as very singular that, with a moderately strong breeze from the north, the barometer should have stood so high, remaining through the day at about 30·3 inches, and marking at nine p.m. 30·28. The temperature, as was to be expected, was higher than on the previous day, being about 40° during the day, and not falling at night below 35°.

Although the morning showed some improvement in the appearance of the weather, the sky was gloomy when, after a little trouble in raising the anchor, we got under way early on the 7th of June. The clouds lifted occasionally during the day, and I enjoyed some brief glimpses of grand scenery; but the only distinct impression I retained was that of hopeless bewilderment in attempting to make out the positions of the endless labyrinth of islands through which we threaded our way. In spite of all that has been done, it seems as if there remained the work of many surveying expeditions to complete the exploration of these coasts. As to several of the eminences that lie on the eastern side of the channel, it is yet uncertain whether they are islands or peninsulas projecting from the mainland. It was announced that our next anchorage was to be at Puerto Bueno, there being no other suitable place for a considerable distance, and we were led to expect that we should probably find there some Fuegians, as the place is known to be one of their favourite haunts.

We dropped anchor about half-past two, in a rather wide cove, or small bay, opening into the mainland a few miles south of Chatham Island. The shores are comparatively low, and enclosed by a dense forest of evergreen beech, which in most parts descends to the water's edge. The place owes its good repute among mariners to the excellent holding-ground; but it did not appear to me as well sheltered as the other natural harbours that we visited, and as the bottom shelves very gradually, we lay fully a mile off the shore. Fortunately the weather had improved somewhat; a moderate breeze from the north brought slight drizzling rain, but gave no further trouble. A boat was soon ready alongside, and we pulled for the shore, with three of the ship's officers armed with fowling-pieces, intended partly to impress the natives with due respect, but mainly designed for the water-birds that abound along the shores of the inlet. We were correctly steered for the right spot, as, on scrambling ashore and crossing the belt of spongy ground between the water and the edge of the forest, we found evident tokens that the Fuegian encampment had not been long deserted. The broken remains of a rude canoe and fragments of basket-work were all that we could find, and we judged that a small party, perhaps no more than ten or a dozen, had left the place a few weeks before our arrival. These wretched Fuegians are said to go farther south, and to keep more to the exposed coasts during winter, because at that season animal life is there more abundant.

After exchanging sundry jokes about the general disappointment in failing to behold the *wilde fräulein* in their natural home, the party separated, two of the officers proceeding in the boat towards the upper part of the inlet in quest of water-fowl. For nearly an hour we heard the frequent discharge of their guns, and much ammunition must certainly have been expended; but when they returned their report was that the birds were too wild, and no addition was made to the ship's larder.

The general character of the vegetation at Puerto Bueno was the same as that at Eden Harbour, but there were some indications of a slight increase in the severity of the climate. *Mitraria coccinea* and a few other representatives of the special flora of Chili were no longer to be found, while some antarctic types not before seen here first made their appearance. The most prominent of these was a bush from three to five feet high, in general appearance reminding one of rosemary, but at this season abundantly furnished with the plumed fruits characteristic of a composite. This plant, nearly allied to the genus *Olearia*, whose numerous species are confined to Australia, New Zealand, and the adjoining islands, is known to botanists as *Chiliotrichium amelloides*, and is one of the characteristic species of this region. It is plentiful in Fuegia and on the northern shores of the Straits of Magellan. Sir Joseph Hooker, in the "Flora Antarctica," remarks that this is the nearest approach to a tree that is made by the meagre native vegetation of the Falkland Islands.

My attention had already been directed at Eden Harbour to the peculiar coniferous plants of this region, and I here found the same species in better condition. The most conspicuous, a small tree with stiff pointed leaves somewhat like an araucaria, here produced abundant fruit, which showed it to be a *Podocarpus* (*P. nubigena* of Lindley). Another shrub of the same family, but very different in appearance, is a species of *Libocedrus*, allied to the cypress of the Old World, which tolerates even the inclement climate of Hermite Island, near Cape Horn. The distribution of the various species of this genus is not a little perplexing to the botanical geographer. This and another species inhabit the west side of South America, two are found in New Zealand, one in the island of New Caledonia, one is peculiar to Southern China, and one to Japan, while an eighth species belongs to California. The most probable supposition is that the home of the common ancestor of the genus was in the circumpolar lands of the Antarctic Circle at a remote period when that region enjoyed a temperate climate; but the processes by which descendants from that stock reached such remote parts of the earth are not easily conjectured.

It was nearly dark when the unsuccessful sportsmen returned with the boat, and but for the ship's lights we should have scarcely been able to make out her position. Some of the many stories of seamen cast away in this inclement region came into my mind during the short half-hour of our return, and, in the presence of the actual scenes and conditions, my impressions assumed a vividness that they had never acquired when "living at home at ease."

In the evening I observed that the barometer had fallen considerably from the usually high point at which it stood up to the 6th, and throughout the night and the following day (June 8) it varied little from 29·9 inches. When we came on deck on the morning of the 8th, the uniform remark of the passengers was, "What a warm day!" We had become used to a temperature of about 40°, and a rise of 5° Fahr. gave the impression of a complete change of climate. It is curious how completely relative are the impressions of heat and cold on the human body, and how difficult it is, even for persons accustomed to compare their sensations with the instrument, to form a moderately good estimate of the actual temperature. We paid dearly, however, for any bodily comfort gained from the comparative warmth in the thick weather that prevailed during most of the day. We had some momentary views of grand scenery, but, as on the preceding day, these were fleeting, and I failed to carry away any definite pictures. It would appear that in such weather the navigation amid such a complete maze of islands and channels must be nearly impossible, but the various surveying-expeditions have placed landmarks, in the shape of wooden posts and crosses, that suffice to the practised eyes of seamen.

About ten a.m. we reached the end of the Sarmiento Channel, opposite to which the comparatively broad opening of Lord Nelson Strait, between Hanover Island and Queen Adelaide Island, leads westward to the Pacific, and before long entered on the third stage of our voyage, which is known as Smyth's Channel. This name is used collectively for the labyrinth of passages lying among the smaller islands that fill the space between Queen Adelaide Island and the mainland of South-western Patagonia; but to distinguish the openings between separate islands various names have been given, with which no one not a navigator need burthen his memory. Perhaps the thick weather may have been the cause, but we all noticed the comparative rarity of all appearance of animal life on this and the previous day. A large whale passing near the ship gave the only occasion for a little momentary excitement. As we ran southward, and were daily approaching the winter solstice, the successive days became sensibly shorter, and it was already nearly dark when, soon after four p.m., we cast anchor in an opening between two low islands which is known as Mayne Channel.

It was impossible not to experience a sense of depression at the persistence of such unfriendly weather during the brief period of passing through a region of such exceptional interest, an opportunity, if once lost, never to be recovered. With corresponding eagerness the hope held out by a steady rise of the barometer was greeted, especially when I found that this continued up to ten p.m., and amounted since morning to a quarter of an inch. We were under way some time before daylight on June 9, and great was my delight when, going on deck, I found a cloudless sky and the Southern Cross standing high in the firmament.

It was a morning never to be forgotten. We rapidly made our way from amid the maze of smaller islands, and glided over the smooth water into a broad channel commanding a wide horizon, bounded a panorama of unique character. As the stars faded and daylight stole over the scene, fresh features of strangeness and beauty at each successive moment came into view, until at last the full glory of sunshine struck the highest point of Queen Adelaide Island, and a few moments later crowned the glistening summits of all the eminences that circled around. The mountainous outline of Queen Adelaide Island, on the right hand, which anywhere else would fix attention, was somewhat dwarfed by the superior attractions of the other objects in view. We had reached the point where Smyth's Channel widens out into the western end of the Straits of Magellan, and right in front of us rose the fantastic outline of the Land of Desolation, as the early navigators styled the shores that bound the southern entrance to the Straits; and as we advanced it was possible to follow every detail of the outline, even to the bold summit of Cape Pillar, forty miles away to the westward. Marking as it does the entrance to the Straits from the South Pacific, that headland has drawn to it many an anxious gaze since steam navigation has made the passage of the Straits easy and safe, and thus avoids the hardship and delay of the inclement voyage round Cape Horn.

The coast nearest to us was at least as attractive as any other part of the panorama. The southern extremity of the continent is a strange medley of mountain and salt water, which can be explained only by the irregular action of elevatory forces not following a definite line of direction. Several of the narrow sounds that penetrate the coast are spread out inland into large salt-water lakes, and all the shores along which we coasted between Smyth's Channel and Sandy Point belong to peninsulas projecting between fifty and one hundred miles from the continuous mainland of Patagonia. The outline is strangely varied. Bold snow-covered peaks alternate with lower rocky shores, and are divided by channels of dark blue water penetrating to an unknown distance into the interior. From amidst the higher summits flowed several large ice-streams, appearing, even from a distance, to be traversed by broad crevasses. I did not see any of these glaciers actually reach the sea, but one, whose lower end was masked by a projecting forest-clad headland, must have approached very near to the beach.

I have called the scene unique, and, in truth, I believe that nothing like it is to be found elsewhere in the world. The distant picture showing against the sky under the low rays of the winter sun is probably to be matched by some that arctic navigators bear in their memory; but here, below the zone of snow and ice, we had the striking contrast of shores covered by dense forest and clothed with luxuriant vegetation. Not much snow can have fallen, as up to a height of about twelve hundred feet above the sea, as far as the forest prevails, none met the eye. On the Norwegian coast, where one might be tempted to look for winter scenes somewhat of the same character, the forest is composed of coniferous trees, which have a very different aspect, and at the corresponding season they are, I imagine, usually so laden with snow that they can give little relief to the eye.

I was struck by the fact that, although we had travelled southward five and a half degrees of latitude (nearly four hundred English miles) since entering the Gulf of Peñas, the upper limit of the forest belt was so little depressed. I could not estimate the average depression at more than from two to three hundred feet.

As we advanced into the main channel, and were drawing near to the headland of Cape Tamar, where the Straits of Magellan are narrowed between that and the opposite coast of the Land of Desolation, we noticed that what seemed from a distance to be a mere film of vapour lying on the surface of the sea grew gradually thicker, rose to a height of about one hundred feet, and quite abruptly, in the space of two or three ship's lengths, we lost the bright sky and the wonderful panorama, and were plunged in a fog that lasted through the greater part of the afternoon. The one constant characteristic of the climate of this region is its liability at all seasons to frequent and abrupt change, especially by day. It is, as I learned, a rare event when a day passes without one or two, or even more frequent, changes of the wind, bringing corresponding changes of temperature, rain, or snow, or clear sky; but, as a rule, the weather is less inconstant in winter than at other seasons. A short experience makes it easy to understand the extreme difficulty of navigation in the Straits for sailing ships, and the expediency of preferring the less inviting course of rounding Cape Horn.

Several times during the day the fog cleared away for a while, and gave us grand views of the coast on either hand. That of the Land of Desolation especially attracted my attention. Captain Willsen pointed out to me, as we stood on the bridge, to which I had free access, the opening of a narrow sound which has lately been ascertained to penetrate entirely through what used to be considered a single island. The expressive name must, indeed, be abandoned, for, if I am not mistaken, the Land of Desolation of our maps is already known to consist of three, and may possibly form many more islands, divided from each other by very narrow channels. Our cautious commander resolved once again to anchor for the night, and selected for the purpose Borya Bay, a small sheltered cove some distance east of Port Gallant, a harbour often visited by the English surveying-expeditions. Daylight had departed when, about half-past five, we reached our anchorage; but the sky was again quite clear, and we enjoyed the weird effects of moonlight illumination. The scenery is very grand, and was more wintry in aspect than at any other point in our voyage. A mountain at the head of the cove rose steeply to a height of at least two thousand feet, and cast a dark shadow over the ship as we lay very near the shore. The shores were begirt with the usual belt of forest, but this did not extend far, and the declivities all around were clad with snow, which lay rather deep. It appeared to me that a rather large glacier descended to within a few hundred feet of the shore, but, seen by the imperfect light, I felt uncertain as to the fact. Since entering the Straits, I had noticed that on the steeper slopes facing the south, where the surface can receive but little sunshine at any season, the forest ascends but a short distance above the sea-level. Above that limit in such situations I observed only a scanty covering of bushes, and higher up the surface at this season appeared quite bare.

As Borya Bay is one of the customary haunts of the Fuegians, the steam-whistle was sounded on our arrival as an invitation to any natives who might be encamped there. This always suffices to attract them, with the hope of being able to gratify their universal craving for tobacco. The appeal was not answered, as the people were doubtless on the outer coasts, and we were not destined to see anything of the most miserable of all the races of man.

As the weather remained bright, the anchor was raised soon after midnight, and by one a.m. we were on our way, steering south-east, to round the southern extremity of the mainland of America. Awaking to the disappointment of having missed a view of one of the most interesting portions of the Straits, I hurried on deck, and found a new change in the aspect of the skies. The night had been cold, with a sharp frost; but in the morning, soon after daybreak, the air felt quite warm, with the thermometer marking 39° Fahr. A northerly breeze had set in, and as an inevitable result brought thick weather. I again noticed, however, that the barometer on these coasts seems to be very slightly affected by changes in the wind's direction. It stood last night at 30·16 inches, and on the morning of the 10th, with a complete change of weather, had fallen only eight-hundredths of an inch.

The southern end of the continent is shaped like a broad wedge, whose apex is Cape Froward, laying in south latitude 53° 54'. We passed it early in the forenoon, giving the headland, which we saw dimly to the north, a broad berth, so that we about touched the 54th parallel. If we compare this with the climate of places in about the same latitude, as, for instance, with that of the Isle of Man, we are apt to consider the climate as severe; but we habitually forget how far the condition of Western Europe is affected by exceptional circumstances; and if we look elsewhere in the northern hemisphere, taking, for instance, the Labrador coast, the south of Kamschatka, or even the coast of British Columbia, we must admit that the Straits of Magellan afford no confirmation to the prevalent ideas respecting the greater cold of the climate of the southern hemisphere.

Soon after this turning-point of the voyage the sky partially cleared to the southward, and we were fortunate enough to enjoy one of the most impressive scenes that my memory has recorded. The broad sound that divides Clarence Island from the main island of Tierra del Fuego lay open before us, flanked on either hand by lofty snow-clad summits. In the background, set as in a frame, rose the magnificent peak of Mount Sarmiento, the Matterhorn of this region, springing, as it appeared, from the shore to a height of seven thousand feet.* Sole sovereign of these antarctic solitudes, I know of no other peak that impresses the mind so deeply with the sense of wonder and awe. As seen from the north, the eastern and western faces are almost equally precipitous, and the broad top is jagged by sharp teeth, of which the two outermost, one to the east, the other to the west, present summits of apparently equal height. At a distance of about twenty-five miles the whole mass seemed to be coated with snow and ice, save where some sharp ridges and teeth of black rock stood out against the sky. I remained for some time utterly engrossed by the marvellous spectacle, and at last bethought myself of endeavouring to secure at least an outline of the scene; but before I could fetch a sketch-book, a fresh change in the weather partly obscured, and, a few minutes later, finally concealed from my eyes a picture that remains vividly impressed on my memory.

It was impossible not to speculate on the origin and past history of this remarkable peak. Admitting that there is evidence to show that the larger part of the rocks of this region are of volcanic origin, it appeared to me evident not only that Mount Sarmiento is not a volcanic cone, but that the rock of which it is composed is not of volcanic origin. Whether its real form be that of a tower, or that of a ridge with precipitous sides seen in profile, no volcanic rocks elsewhere in the world can retain slopes so

---

* I am not aware that the concurrent conclusions as to the height of this mountain have been verified by accurate observations, but the height commonly given appears to be a close approximation to the truth.

nearly approaching to the vertical. It is, I believe, a portion of the original rock skeleton that formed the axis of the Andean chain during the long ages that preceded the great volcanic outbursts that have covered over the framework of the western side of South America. Like most peaks of a similar form, I am disposed to believe that in the course of gradual upheaval the flanks have been carved by marine action to the nearly vertical form which impresses the beholder. Although snow-covered mountains suffer a certain limited amount of denudation in the channels through which glaciers flow, there is reason to hold that they are far less subject to degradation than those which are not protected from the main agencies that wear away rocky surfaces. It is by alternations of temperature, by frost, and the action of running water, that rocks are rapidly eaten away, and from these a snow-covered mountain is to a great extent secured.

A few miles east of Cape Froward the coast of the mainland trends nearly due north for a distance of fully sixty miles, and a marked change is perceived in the aspect of the shores. Instead of the bold outlines to which our eyes had become accustomed, the coast-line lay low, fringed with forest on the side of the mainland, which now lay to our west, and on the other hand showing bare flats, here and there flecked with fresh snow. The land on that side at first belonged to Dawson Island; but later in the day, as we approached our destination, the dreary flats formed part of Northern Tierra del Fuego.

The weather was thick as we passed Port Famine,

and there was little to attract attention until we drew near to Sandy Point, a place that was to me the more interesting as I intended to make it my home until the arrival of the next English steamer. The belt of forest rose over low swelling hills near the sea, and in the distance a loftier range, from two to three thousand feet in height, showed a nearly horizontal line against the cloudy sky. As we approached, several structures of painted wood became visible, and for the first time since we left Lota we beheld human dwellings. Sandy Point, known to the natives of South America by the equivalent name Punta Arenas, is certainly one of the most isolated of inhabited spots to be found in the world. Since the scramble for Africa has set in, it is, I suppose, only on the Australian coast that one would find any settlement so far removed from neighbours or rivals. On the side of Chili the nearest permanent habitations are in the island of Chiloe, fully seven hundred miles distant in a straight line, and considerably farther by the only practicable route. On the side of Argentaria there is a miserable attempt at a settlement at the mouth of the river Santa Cruz, where the Argentine Government has thought it expedient to hoist their flag in order to assert the rights of sovereignty of the Confederation over the dreary wastes of South-eastern Patagonia. This was described to me as a group of half a dozen wooden sheds, where a few disconsolate soldiers spend a weary time of exile from the genial climate of Buenos Ayres. By the sea route it is about four hundred miles from Sandy Point, but no direct communication between the two places is kept up. For all practical purposes, the nearest civilized neighbours to Sandy Point are the English colonists in the Falkland Islands, where, in spite of inhospitable soil and climate, some of our countrymen have managed to attain to tolerable prosperity, chiefly by sheep-farming. But with an interval of nearly five hundred miles of stormy ocean mutual intercourse is neither easy nor frequent.

CHAPTER V.

Arrival at Sandy Point—Difficulties as to lodging—Story of the mutiny—Patagonian ladies—Agreeable society in the Straits of Magellan—Winter aspect of the flora—Patagonians and Fuegians—Habits of the South American ostrich—Waiting for the steamer—Departure—Climate of the Straits and of the southern hemisphere—Voyage to Monte Video—Saturnalia of children—City of Monte Video—Signor Bartolomeo Bossi; his explorations—Neighbourhood of the city—Uruguayan politics—River steamer—Excursion to Paisandu—Voyage on the Uruguay—Use of the telephone—Excursion to the camp—Aspect of the flora—Arrival at Buenos Ayres—Industrial Exhibition—Argentine forests—The cathedral of Buenos Ayres—Excursion to La Boca—Argentaria as a field for emigration.

THE time had come for parting with my genial fellow-traveller, Mr. H——, with our excellent captain, and with the officers of the *Rhamses*, to all of whom I felt indebted for friendly aid in my pursuits; and on entering the boat that was to take me ashore I was introduced to the captain of the port, an important official of German origin. Of his various excellent qualities, the only one that I at first detected was a remarkable gift of taciturnity, rarely interrupted by a single monosyllable. I was aware that accommodation for strangers at Sandy Point is extremely limited, but I consoled myself with a belief that, if it came to the worst, the letter which I carried to the governor from the minister for foreign affairs at Santiago would help me through any preliminary difficulties. On reaching the shore, my luggage was without further question carried to a house close by, which is at this place the sole representative of a hotel. The accommodation available for strangers consists of a single room of fair dimensions, and this, as I soon learned, was occupied by a stranger. A glance at the multitudinous objects scattered about made me feel sure that the visitor must be a brother naturalist, but did not help me to solve the immediate difficulty. As I stood at the entrance, a dark-haired person, speaking pretty good English, proposed to take me to the house of the English vice-consul, and in his company I had the first view of the settlement of Sandy Point. As the ground rises very gently from the beach, few houses are seen from the sea, and the place is not so inconsiderable as it at first appears. Though rather to be counted as a village than as a town, it has the essential privilege of a Spanish city in the possession of a *plaza*, not yet quite surrounded by houses. The buildings are small, and nearly all built of wood painted outside.

The next piece of information received was unfavourable to my prospects. An Argentine corvette had reached Sandy Point a few days before, and the vice-consul had been invited, along with the governor and other notabilities, to a luncheon, which was likely to last for some time. I was fortunately provided with a note of introduction to Dr. Fenton, the medical officer of the settlement, which I now proceeded to deliver. Being somewhat unwell, he had not joined the marine entertainment, and I was at once cordially received. Not many minutes were needed to discover in my host a fellow-countryman, one of a family in the county of Sligo, with which I had some former acquaintance. Possessing in large measure the national virtue of hospitality, Dr. Fenton might have perhaps been satisfied with even a slighter claim; but, as it was, I from that time continued during my stay to receive from him the utmost kindness and attention. The first short conversation made me much better acquainted with the history of the settlement than I was before my arrival.

In 1843 the Chilian Government decided on establishing a penal settlement in the Straits of Magellan, and selected for its position Port Famine, which had been frequently visited by early navigators. After a few years' experience that place was abandoned, and the settlement was transferred to Sandy Point. This was partly preferred on account of a deposit of lignite of inferior quality, which lies little more than a mile from the shore. A considerable number of convicts were maintained at the station, and as there was little risk of escape they were allowed considerable liberty. At length, in 1877, the injudicious severity of the governor of that day provoked a revolt among the convicts. They speedily overcame the keepers, and the officials and peaceable inhabitants had no resource left but to fly to the forest. The convicts proceeded to set fire to the houses. Dr. Fenton lost his house, furniture, and books, and, in addition, the record of ten years' meteorological observations. By a fortunate accident, a Chilian war-vessel reached Sandy Point just when disorder was at its height ; the insurgents were speedily overpowered, and several of the ringleaders executed. The weather was unusually mild, and the refugees, amongst whom were many ladies and young children, suffered less than might have been expected in such a climate. Nearly all the houses seen by me had been hastily erected since the outbreak, and, as was natural, were on a scale barely sufficing for the wants of the inmates.

I fully understood that no amount of hospitable intentions could enable Dr. Fenton to give me quarters in his house, and he assured me that the governor, Don Francisco Sampayo, was no less restricted as to accommodation. One resource, however, seemed available : the German consul, Herr Meidell, had returned for a visit to Europe, and it was thought that, on application to his partner, a room might certainly be obtained in his house. My dark-haired friend, who had reappeared on the scene, and who turned out to be a native of Gibraltar, kindly undertook to arrange the matter, and, after an early dinner at Dr. Fenton's hospitable table, I proceeded with him to present my letter to the governor. The great man had not yet returned to shore, but I made the acquaintance of his wife, a delicate Peruvian lady, who sat, wrapped in a woollen shawl, in a room without a fire, of which the temperature must have been about 45° Fahr. On leaving the governor's house, we again encountered my envoy, whose countenance at once proclaimed that he had failed in his mission. Mr. Meidell, being a cautious man, had locked up most of his furniture and household effects before going to Europe, and had left strict injunctions that no one was to enter the part of his house used as a private dwelling. As we stood consulting about further proceedings, a tall figure approached, and I learned that it belonged to the stranger who occupied the solitary room available for visitors to Sandy Point.

I speedily made the acquaintance of Signor Vinciguerra, one of the group of energetic young Italian naturalists whose head-quarters are at Genoa. He belonged to the expedition commanded by Lieutenant Bove of the Italian navy, and had remained at Sandy Point to investigate the zoology of the neighbouring coast, while his companions proceeded to Staten Island, or Isla de los Estados, at the eastern extremity of the Fuegian Archipelago. Community of pursuits and several mutual friends at once cemented cordial relations, and Signor Vinciguerra kindly undertook to make room for me in his rather restricted quarters. We proceeded to the house close by the landing-place, and I was in the act of arranging the matter with the landlord, when the British vice-consul appeared. He had overcome the scruples of Mr. Meidell's partner, a mattress and some coverings had been found, a room was at my disposal, with a bed on the floor, and the lodging difficulty was solved.

Not without some regret at being separated from an agreeable companion, I accepted the offered quarters, and had the needful portion of my luggage carried to my temporary home. As the sun set before four o'clock, it was already dark before I was installed in my new quarters, and the evening was spent under the hospitable roof of Dr. Fenton, from whom I received much interesting information as to the region which he has made his home, and the indigenous population. On my way to his house I saw the first specimens of the Patagonian Indians, who at this season frequent the settlement to dispose of skins, chiefly *guanaco* and *rhea*, and indulge in their ruling passion for ardent spirits. Two ladies of large and stout build, attired in shabby and torn European dress, and both far gone in intoxication, were standing at a door of a shop or store, and indulging in loud talk for the entertainment of a circle of bystanders. The language was, I presume, their native dialect, with here and there a word of Spanish or English, and the subject seemed to be what with us would be called chaff, as their remarks elicited frequent peals of laughter. I was suddenly reminded of a drunken Irish basket-woman whose freaks had been the cause of mingled alarm and amusement in my early childhood.

During the day the streets of Punta Arenas were deep in mud, but as I went home at night, the sky was cloudless, a sharp frost had set in, and the mud was hard frozen. I had not before enjoyed so fine a view of the southern heavens. The cross was brilliant, nearly in the zenith, and I made out clearly the dark starless spaces that have been named the coal-sacks.

I was on foot before daylight on the 11th of June. The benevolent German who managed Mr. Meidell's establishment sent up a cup of hot coffee, and a brazier with charcoal, which was grievously wanted to dry my plant-paper. The sky was still clear, and the sun, rising blood-red over the flat shores of Tierra del Fuego on the opposite side of the Straits, was a striking spectacle. I had arranged overnight to take with me a boy having some knowledge of the neighbourhood, and was just starting for a walk when I met the governor, who at this early hour was on his way to call upon me. After a short conversation with this courteous gentleman, and accepting an invitation to dine at his house, I pursued my course in the direction of the now disused coal mine. For about half a mile I followed the tramway which was erected some years ago to carry the coal to the port. It runs along the low ground between the hills and the shore, and then enters a little flat-bottomed valley between the hills. Heavy rain had recently fallen, and the flat had been flooded, but the surface was now frozen over. Before long we found the tramway impracticable; it had been allowed to fall to decay, and, being supported on trestles, the gaps were inconveniently frequent. I then attempted to continue my walk over the flat, and found the ice in some places strong enough to bear my weight, but it frequently gave way, and I soon got tired of splashing through the surface into the ice-cold water, and resolved to betake myself to the adjoining hills. The weather showed itself as changeable on this day as it usually is in this singular climate. For about half an hour the sky was clear and the sun so warm that I could not bear an overcoat. Then a breeze sprung up from the north-west, the sky was soon covered, and some rain fell; again the sky cleared, and, if I remember right, four or five similar changes occurred before nightfall.

At this season I could not expect to see much of the vegetation of the country, but I found rather more than I expected. Two *Compositæ*, both evergreen shrubs, were abundantly clothed with fruit, and among other characteristic forms I collected two species of *Acæna*, a genus widely spread through the southern hemisphere, allied to, but very distinct from, our common *Alchemilla*. From its ancestral home in south polar lands, many descendants have reached South America, and some of these have followed the Andean chain, and thus got to Mexico and California. From the same stock we find representatives in New Zealand, Australia, Tristan d'Acunha, and South Africa, while one has travelled so far as the Sandwich Islands. The seeds are provided with hooked beaks, which may have attached themselves to the plumage of oceanic birds, and a single successful transport in the course of many ages may have introduced the parent of the existing species to new regions of the earth. It was not without interest to find two cosmopolitan weeds, our common shepherd's purse and chickweed, both flowering in winter in this remote part of the world.

From the summit of the hill I enjoyed a good view of the flat-topped range—apparently from 2500 to 3000 feet in height—that separates the Straits of Magellan from Otway Water. This is a land-locked basin nearly fifty miles long and half as wide, connected with the sea by a narrow sound that opens on the western side of the Straits near Port Gallant. The lower slopes of the intervening range are covered with forest, and the summit apparently bare, but in this season covered with snow. If the extreme difficulty of penetrating the forests were not well known, it would be a matter of surprise that no one has ever crossed the range, and that the eastern shores of Otway Water, not thirty miles distant from Punta Arenas, are yet unexplored.

In returning to Punta Arenas I passed through the remains of the burnt forest that once extended close to the houses. In the summer of 1873, either by design or accident, fire seized the forest, composed of large trees of the antarctic beech, and raged so furiously for a time as to threaten destruction to the entire place. After the first efforts at averting the immediate danger, no further interference was attempted, and I was assured that the conflagration was not entirely exhausted until the ensuing winter, nearly six months after it commenced. I passed the charred remains of hundreds of thick stumps, many of them over three feet in diameter, but I was surprised to find several trees much too large to have grown up since the fire, which in some unexplained way escaped destruction. Unlike most of the beeches of the southern hemisphere, this has deciduous leaves, so that the branches were bare; but many of them were laden with the curious parasite, *Myzodendron punctulatum*, the structure of which plant and its allies was long ago admirably illustrated by Sir Joseph Hooker.*

* "Flora Antarctica," vol. ii. p. 289.

The evening of this day was very agreeably spent at the house of the governor, who had invited to his table Commander Pietrabona and two officers of the Argentine corvette, Cabo de Ornos, Signor Vinciguerra, the captain of the port, and two or three of the principal inhabitants. One of the favourable features by which a stranger is impressed in Chili is the comparative moderation with which political conflicts are conducted. In the other South American republics a conspicuous party leader is marked by the opposite party for relentless proscription, and not rarely for assassination. In Chili political offences are condoned. Don Francisco Sampayo, who is a courteous and accomplished gentleman, had been mixed up in the same abortive movements in which Don B. Vicuña Mackenna was concerned, and had with that gentleman undergone a term of exile, but was subsequently appointed by his political opponents to the government of this settlement.

The government house was unpretending, and could not by any stretch of language be called luxurious. Two good reception-rooms and the bedrooms of the family, all on the ground floor, opened into a small court exposed to rain and snow. The reception-rooms had fireplaces, but these were used only in the evenings, and it was not surprising that the governor's wife, brought up in the tepid climate of Peru, seemed unable to resist the inclemency of this region. Their children, however, were vigorous and thriving, reminding one more of English boys and girls than any I had seen in South America. The most interesting figure in the family group was that of the mother of Madame

Sampayo, an elderly lady, with the remains of remarkable beauty, and an unusual combination of dignity and grace with lively, almost playful, conversation. The removal to this inhospitable shore had not quenched her' activity, and she employed her leisure in devising pretty ornaments from seaweeds, shells, and other natural productions of the place.

The Chilian and Argentine Republics concluded, in the year before my visit, a convention to regulate their rival pretensions to the possession of the territory on both sides of the Straits of Magellan, which at one time threatened to engage the two states in war for a worthless object. The new boundary-line is drawn along the middle of the peninsula, ending in Cape Virgenes at the eastern entrance of the Straits, thus leaving to Chili the whole of the northern shores. Opposite to Cape Virgenes is a headland named Cape Espiritu Santo on the main island of Tierra del Fuego. The boundary runs due south from that point, cutting the island into two nearly equal parts, of which the eastern half, along with Staten Island, is assigned to Argentaria. As I understood from the conversation at dinner, Commander Pietrabona had obtained from his government a grant or lease of Staten Island, but it seems very doubtful whether any profit can be derived from an island lying nearly three degrees further south than the Falklands, and fully exposed to the antarctic current.

Amongst the various nationalities that met on this evening, the representative of Germany, the captain of the port, was perhaps the most typical. He is believed to have a more complete and accurate knowledge of the coasts of the Straits of Magellan and of the Channels of Patagonia than any other living man. The conversation was animated, and not seldom turned on the topography of this region; but the worthy Teuton sat obstinately silent, or, when directly appealed to, generally answered by a single monosyllable of assent or negation. A superficial observer would have set this down as evidence of a surly or misanthropic disposition, but in truth this worthy man is noted for good nature and a ready disposition to oblige his neighbours. Having accepted the governor's offer of a horse for an excursion on the following day, I departed with the other guests, and once again enjoyed the view of the southern heavens undefiled by a single cloud, and found the mud of the streets frozen hard.

The dawn of June 12 was again cloudless, and the circle of the red sun, distorted by refraction, rose over the flats of Tierra del Fuego. But in less than a quarter of an hour heavy leaden clouds gathered from all sides and portended a stormy day. I felt rather unwell, and resolved to postpone my intended excursion to the following day. After the needful care given to my plant-collections, I repaired to the hospitable sitting-room of Dr. Fenton, which was, I believe, the only moderately warm spot at Punta Arenas, and passed the day in his company, or that of Mrs. Fenton and their pretty and intelligent children. The heavy rain which persisted nearly all day diminished my regret at having to remain indoors. I made a few notes of the varied information which I obtained from a gentleman who has had unusual opportunities for acquiring knowledge, and who, although not a professed naturalist, appears to be an accurate observer.

The Patagonian Indians who frequent Punta Arenas to dispose of skins appear to be rapidly diminishing in numbers, and one good observer believes that they are now to be counted rather by hundreds than by thousands. The chief cause is doubtless the destructive effect of ardent spirits. They commonly expend nearly everything they gain in drink, but after recovering from a fit of beastly intoxication they usually invest whatever money remains in English biscuits, which they carry off to the interior. Here, as well as at many other places in South America, I heard curious stories showing the extraordinary estimation in which Messrs. Huntley and Palmer are held by the native population. Among the curious customs of these Indians, Dr. Fenton told me that as soon as a child is born one or more horses are assigned to it as property, and if the child should die, as they often do, prematurely, the horses are killed. He further says that a childless Indian not rarely adopts a dog, the ceremony being marked by assigning horses to the dog as his property, and that, as in the case of the human child, at the dog's death the horses are killed.

Agreeing with most of those who have observed the Fuegians in their native home, Dr. Fenton is sceptical as to the possibility of raising that hapless tribe above their present condition. All honour is due to the devoted men who have laboured at the mission station at Ushuaia in the Beagle Channel, and it may be that some partial success has been obtained with children taken at an early age. But, looking around at the multiplied needs of so many other less degraded branches of our race, one is tempted to believe that such noble efforts might more usefully be bestowed elsewhere. Dr. Fenton thinks that the fact, which appears to be well attested, that Fuegians, in a rough sea, when in danger in their frail canoes, have been known to throw an infant overboard, is evidence that they believe in spirits, the child being offered to appease the wrath of supernatural powers. I confess that I place little reliance on the conclusions of civilized men as to the ideas or motives of savage races in a condition so low that we have the most imperfect means of communicating with them.

I was not able to ascertain positively whether the species of *rhea*, or South American ostrich, found near the Straits of Magellan, is exclusively the smaller species (*Rhea Darwinii*), but I believe there is no doubt that the larger bird does not range so far as Southern Patagonia. Dr. Fenton has had frequent opportunities for observing the habits of the bird. He finds that the nests are constructed by the female birds, three or four of these joining for the purpose. One of them deposits a single egg in a hollow place, and over this the nest is built. Each of the females deposits several eggs in the nest, and then wanders away, the male bird sitting on the nest till the young birds are hatched. When this happens the parent clears away the nest, breaks up the egg which lay beneath it, and gives it to the young birds for food. The flesh is described as delicious, somewhat intermediate in flavour between hare and grouse.

Dr. Fenton had commenced the trial of an experiment which, if successful, may hereafter attract settlers to the eastern shores of the Straits of Magellan. The appearance of the country had already shown to me that the climate is much drier here than on the western side of Cape Froward, and I believe that the range above spoken of, which divides this coast from Otway Water, is about the eastern limit of the extension of the zone of continuous forests that cover all but the higher levels of Western Patagonia. Between Peckett Harbour, about forty miles north of Punta Arenas, and the Atlantic coast the country is open and produces an abundance of coarse herbage. Sheep are known to thrive in the Falkland Islands, about the same latitude, and Dr. Fenton had recently procured from that place a flock which he had established in the neighbourhood of Peckett Harbour.

I was warned that the English steamer might possibly arrive in the afternoon of June 13, though more probably on the following day, so that it was expedient to start early on the short excursion which I proposed to make along the coast to the north of Punta Arenas. The horses were ready soon after sunrise, and the governor's secretary was good enough to accompany me. After fording the stream which flows by the settlement, we for some distance followed the sandy beach, dismounting here and there to examine the vegetation. Few plants could at this season be found in a state in which they could be certainly identified, but there was quite enough to reward a naturalist. It was very interesting to find here several cosmopolitan species whose diffusion cannot, I think, be set down to the agency of man.

Of these I may reckon *Plantago maritima*, and a slight variety of our common sea-pink (*Armeria maritima*, var. *andina*). To these I am disposed to add *Rumex acetosella*, which I found creeping in the sand far from the settlement, and a form of the common dandelion (*Taraxacum lævigatum* of botanists). Along with these were several representatives of the antarctic flora—a *Colobanthus*, three species of *Acæna*, a *Gunnera*, an *Ourisia*, and several others. Of bushes the most conspicuous are the berberries, of which I found three species. One of these, which I had already seen in the Channels, has leaves like those of a holly, and is appropriately named *Berberis ilicifolia*. Another, which is very common here (*Berberis buxifolia*), has sweet berries, pleasant to the taste; and the third (*B. empetrifolia*) is a dwarf bush, scarcely a foot high, which seems to be confined to the sandy shore. A taller shrub, which I had seen in the Channels as well as in this neighbourhood (*Maytenus Magellanica* of botanists), is called *Leña dura*, and is valued for the hardness of the wood, useful for many small articles. The genus extends throughout South America, but most of the species inhabit tropical Brazil, and we may look on this as the solitary representative of the tropical flora which has reached the southern extremity of the continent.

Having collected whatever was to be found close to the shore, I proposed to strike inland towards the base of the low hills. The country near was a dead flat, and seemed to offer no obstacle. After riding for about a mile over dry ground, we gradually found ourselves in the midst of shallow pools of water, now frozen over. As we advanced progress became more and more difficult. The heavy rain of the preceding day had partially melted the ice. In some places it was strong enough to bear the horses; but it constantly broke under their feet, and they became restive, very naturally objecting to this mode of travelling. After a while, to my surprise, we struck upon a cart track. This, as I soon saw, led to two or three houses inhabited by a few Swiss settlers, who endeavoured to make a living by raising some vegetables for Punta Arenas. The soil appeared to be rich: in this climate few plants can mature fruit or seed, but the more hardy European vegetables thrive sufficiently. Our difficulties were by no means at an end. The cart track was a mass of half-frozen mud, with holes fully two feet deep, into which the horses plunged, until at last it was not easy to persuade them to move in any direction. I dismounted and ascended a hillock some eighty feet above the plain, but on all sides could see no issue from the maze of shallow frozen pools. With some trouble we reached one of the houses, but, in answer to our inquiries, were told that they knew of no better way to Punta Arenas than by the cart track. Apprehending the arrival of the Pacific Company's steamer, and not wishing to remain another fortnight in this remote region, I resolved to return as best we could, and, as always happens, experience enabled both horses and riders to avoid the worst places, so that we got through better than we had expected.

Having made all ready for the possible arrival of the steamer, whose stay is usually very short, I again enjoyed the hospitality of the governor, and once more found myself in the agreeable society of Signor Vinciguerra. One of the many laudable characteristics of Chilian society, in striking contrast with their kinsmen in Spain, is the genuine anxiety commonly shown for the education of the rising generation. It is, indeed, rather amusing to note the tone of contemptuous pity with which the Chilians of pure Spanish descent speak of their European cousins, who are usually denominated "los Gotos." The governor's eldest son had been sent to Germany to pursue his studies, and the services of a young German, who apparently had got into some scrape connected with politics in his own country, had been secured to conduct the education of the younger children. Before dinner the preceptor was engaged in guiding the fingers of one child upon an old pianoforte, and immediately after dinner lessons were resumed with the other children.

In the course of the evening we had a curious illustration of the difficulty of speaking correctly two closely allied dialects. Conversing in Italian with Signor Vinciguerra, a laugh was raised against me for introducing a Spanish word into a sentence; but this was redoubled when, a few minutes later, my Italian friend did exactly the same thing.

Thought is inextricably linked with the impressions derived from the senses, which, excepting with the deaf and dumb, are ordinarily based upon language; and whenever a man speaks with even moderate fluency the fact implies that he thinks in that language. The effort of changing from one language to that of another is that of changing, so to say, the channel through which thought runs. When they are sufficiently different there is no difficulty in maintaining thought within the assigned channel; but when the languages, or dialects, are nearly alike, it is much more difficult to maintain the intended course. It seems to me, indeed, that there is a link of association not only between the idea and the word, but also with the sound of the word. There is comparatively little difficulty in passing from one language to another, though etymologically near akin, when the prevailing sounds are different. Thus, although Portuguese and Spanish are so nearly allied, it is easier to pass from one to the other than from Spanish to Italian, because the phonetic differences are greater in the former case.

The night passed without disturbance, though I had made all ready in case of being summoned to embark; but as the arrival of the steamer was confidently predicted, I completed my arrangements, and removed my luggage to the office of the port captain on the morning of the 14th. The weather was nearly quite dry all day, with a prevailing sharp wind from the south-west, varied by two or three abrupt changes. I did not venture to go into the country, and contented myself with trotting up and down, mainly with the object of keeping myself warm. Evening closed; but no steamer appeared, and I accepted Dr. Fenton's offer of a sofa in his sitting-room for the night, whereon to await the expected summons. Towards four o'clock I sallied forth, without disturbing the household. Profound silence prevailed throughout the settlement; the stars of the southern hemisphere beamed with extraordinary brilliancy, and the muddy streets were iron-bound with frost. After another doze on the sofa, I again went out at dawn, and enjoyed a beautiful sunrise.

The morning of June 15 was unusually favourable for distant views. Beyond the low, bare flats of Tierra del Fuego there showed to the south-east a range of hills, or mountains, whose heights I estimated at from 3500 to 4000 feet, but it is needless to say that, with unfamiliar atmospheric conditions, where the judgment as to distance is so uncertain, such an estimate is quite unreliable. Nearly due south lies Dawson Island, and several high summits were visible in that direction, but I do not believe that either Mount Darwin or Mount Sarmiento are visible from this part of the coast.

During the day I went a short way along the shore to the south, passing the cemetery wherein lie the bodies recovered from the wreck of the *Doterel*. The origin of the explosion which caused that ship to go down with all hands within sight of the settlement, was long a matter of doubt. The most probable opinion is that it was due to the spontaneous ignition of gas generated in unventilated coal-bunkers. Nearly opposite lay the hull of another ship which became a partial wreck on this coast. It contained a cargo of Welsh coal, which is sold at the heavy price of four pounds a ton, and occasionally serves for steamers whose supply has run short.

Along the sandy shores the most conspicuous plant, with large white cottony leaves, is a species of *Senecio*

(*S. candidans* of botanists), which, with nearly twenty others, represents that cosmopolitan genus in this region. What light would be thrown on the past history of the vegetable kingdom if we could learn the origin of that vast genus, and the processes by which it has been diffused throughout the world! Of about nine hundred known species that extend from the Arctic Circle to high southern latitudes, and from the highest zone of the Alps, the Himalayas and the Andes, to the low country of Brazil and the scorching plains of North and South Africa, the great majority are confined to small areas, and are unusually constant in structure, thus presenting a marked contrast to the ordinary rule, dwelt on by Darwin, that among genera that extend over a large portion of earth and have numerous species, the species, or many of them, are themselves widely spread and vary much in form. Neither do we find among the crowd of species many indications of the general tendency to form groups of species nearly allied in appearance and structure within the same geographical area. Many of the very numerous South American species are nearly allied to European and Asiatic forms. Thus in the comparatively small area of Europe we find the representatives of groups characteristic of regions widely separated, and even in the poverty-stricken flora of Britain such different forms as the common groundsel, the ragwort of neglected fields, and the less common *Senecio paludosus*, and *S. campestris*.

The day wore on, and yet no steamer appeared. Knowing people began to speculate on the possibility of some accident having delayed her arrival, or surmised the prevalence of such thick weather about the western entrance to the Straits as might have led her commander to make the circuit by Cape Horn. In the latter case, I should be detained for another fortnight, and although I should have gladly seen something more of the country, and found myself meantime fortunate in pleasant society, I did not in this season desire so long a delay. Once more I betook myself at night to the sofa in Dr. Fenton's hospitable house, and at length, about four in the morning, a tapping at the window announced that the lights of the steamer were in view. Dr. Fenton, who wished to go on board, was speedily ready, and we went to the landing-place where, until the jetée, still in construction, should be finished, the boats are run up on the sandy beach. There was some delay in finding the key of the store where my luggage was housed, but at last we were ready to start. The boat, however, was fast aground on the flat margin of the bay; in vain the four boatmen shoved with their oars, until the taciturn port captain barked out the order to get into the water and shove her off. It was freezing hard, and I fear the poor fellows wished me and my luggage no good when, after much striving, we were finally afloat, and they resumed their places at the oars. In the dark the great hull loomed gigantic as, about five a.m., we pulled alongside of the steamer, which turned out to be the *Iberia*, one of the largest and finest vessels of the Pacific Company, commanded by Captain Shannon.

Having learned that the steamer had been detained by very heavy weather in the South Pacific, and had

had great difficulty in making Cape Pillar, the western landmark of the Straits, I bade farewell to my kind host, and sought for quarters in the great floating hotel. There is something depressing in arriving in a place of entertainment on a cold night, when it is obvious that one's appearance is neither expected nor desired. After a while a steward, scarcely half awake, made his appearance, and arranged my berth. I soon turned in, and slept until near nine o'clock, when we were already well on our way towards the Atlantic opening of the Straits. The morning was bright and not very cold, and for the first time since I entered this region the weather remained unchanged during the day, and the sky clear, with the exception of heavy banks of cloud which showed in the afternoon above the southern and western horizon.

In the morning, when about twenty miles north of Sandy Point, and nearly abreast of Peckett Harbour, the unmistakable peak of Mount Sarmiento was for a short time distinctly seen. It is needless to say that this was due to atmospheric refraction, for the distance was rather over a hundred English miles, and in a non-refracting atmosphere a mountain seven thousand feet high would be below the visible horizon at a distance of about eighty-five miles. Of Mount Darwin, which is believed to be the highest summit of the Fuegian Archipelago, I was not destined to see anything; it is probably completely concealed by the range which runs across the main island of Tierra del Fuego.

The scenery of the eastern side of the Straits of Magellan offers little to attract the eye, the shores on both sides being low and little varied. From Cape Froward to Peckett Harbour the Patagonian coast runs nearly due north, and then trends east-north-east for about seventy miles, where the channel is contracted between the northern shore and Elizabeth Island. After passing the island, we entered the part called "The Narrows," where the Fuegian coast approaches very near to the mainland of the continent. As the day was declining, we issued from this channel into a bay fully thirty miles wide, partly closed by two headlands, which are the landmarks for seamen entering the Straits from the Atlantic. That on the Fuegian side is Cape Espiritu Santo, and the bolder promontory on the northern side is the Cape Virgenes. To a detached rock below the headland English seamen have given the name Dungeness. In the failing light, I could see that the coast westward from Cape Virgenes rises into hills, which appeared to be bare of forest. I should guess their height not to exceed two thousand feet, if it even reaches that limit.

It was almost quite dark when we finally re-entered the Atlantic, and found its waters in a very gentle mood. In these latitudes the name Pacific is not well applied to any part of that which the older navigators more fittingly designated the Southern Ocean.

It was impossible to live for more than a week in winter, at the southern extremity of the American continent, without having one's attention engaged by the singular features of the climate of this region, and especially by their bearing on wider questions which have of late years assumed fresh importance. Mainly through the writings of Dr. James Croll, and the remarkable ability and perseverance with which he has sustained his views, geologists and students of every other branch of natural science have learned to estimate the influence which the secular changes in the eccentricity of the earth's orbit may have exercised on the physical condition of our planet. I have ventured, in the Appendix, to discuss some portions of the vast range of subjects treated of by Dr. Croll,* and to state the reasons which force me to dissent from some of his conclusions ; but I shall here merely say that the impressions derived from my own short experience have been confirmed by subsequent diligent inquiry, and especially by the writings of Dr. Julius Hann, most of which have been published since my return to England.

The belief that the mean temperature of the southern is considerably lower than that of the northern hemisphere was, until recently, prevalent among physical geographers, and has been assumed as an undoubted fact by Dr. Croll. He accounts for it by the predominance of warm ocean currents that pass from the southern to the northern hemisphere within the tropics, and which, as he maintains, ultimately carry a great portion of the heat of the equatorial regions to the north Temperate and Frigid Zones. I think that this belief, as well as many others regarding physical geography, originated in the fact that physical science in its more exact form, had its birth in Western Europe, a region which, especially as to climate, is altogether exceptional in its character. The further our knowledge, yet too limited, has extended in the southern

* See Appendix B.

hemisphere the less ground we find for a belief in the supposed inferiority of its mean temperature. What we do find, in exact conformity with obvious physical principles, is that in the hemisphere where the water surface largely predominates over that of land, the temperature is much more uniform than where the land occupies the larger portion of the surface. In the former, the heat of summer is mainly expended in the work of converting water into vapour, and partially restored in winter in the conversion of vapour into water or ice.

We unfortunately possess but three stations in the southern hemisphere, south of the fiftieth degree of latitude, from which meteorological observations are available, and these are all in the same vicinity—the Falkland Islands, Punta Arenas, and Ushuaja, the mission station in the Beagle Channel at the south side of the main island of Tierra del Fuego. The following table shows the mean temperature of the year at these stations in degrees of Fahrenheit's scale.

	South latitude.	Mean temperature of year.
Falklands	51° 41′	about 43·00°
Punta Arenas	53° 25′	43·52°
Ushuaja	54° 53′	42·39°

If we compare these with the results of observations at places on the east side of continents in the northern hemisphere, we find the latter to show a very much more rigorous climate. Nikolaiewsk, near the mouth of the Amur, in lat. 52° 8′ north, has a mean annual temperature of 32·4° Fahr.; and at Hopedale, in Labrador, lat. 55° 35′, the mean is certainly not higher than 26° Fahr. Even in the island of Anticosti,

T

at the mouth of the St. Lawrence, lat. 49° 24′ north, the main yearly temperature is only 35·8°, or more than 7° below that of the Falkland Islands. But it may be truly said that, although the stations now under discussion are on the eastern side of the South American continent, they virtually enjoy an insular climate, and that there is probably little difference between their temperature and that of places on the west side of the Straits of Magellan.

On comparing the few places out of Europe from which we possess observations in high northern latitudes, I think that the station which admits of the fairest comparison is that of Unalaschka in the North Pacific. The observations at Illiluk in that island, in lat. 53° 53′ north, show a mean annual temperature of only 38·2° Fahr., while at Ushuaja, 1° farther from the equator, the mean temperature is higher by more that 4°. It is true that at Sitka, in lat. 57° north, we find a mean temperature of 43·28° Fahr., or about the same as that of the Falklands. But the position of Sitka is quite exceptional. It is completely removed from the influence of the cold currents that descend through Behring's Straits, and a great mountain range protects it from northerly winds; south-westerly winds prevail throughout the year, and a very heavy rainfall, averaging annually eighty-one inches, imports to the air a large portion of heat derived from equatorial regions. On the coast of Western Patagonia and Southern Chili, this source of heat is partly counteracted by the cold antarctic current that sets along the western coast of South America.

The general conclusion, which seems to be fully established, is that the southern hemisphere is not colder than the northern, and that all arguments based upon an opposite assumption must be set aside.

Among the passengers on board the *Iberia* were a large proportion of ladies and children, the families of English merchants settled in Chili. They had been miserable enough during the three or four days before entering the Straits. The weather had been very severe, and, large as is the vessel, heavy seas constantly broke over her upper deck, so that even the most adventurous were confined to the cabins, very many to their berths. The change to quiet waters and brighter skies acted like a charm, and the spirits of the passengers rose even more than the barometer. The children naturally became irrepressible, and left not a quiet corner in the whole ship. Having first invaded the smoking-cabin and made it the chief *depôt* for their toys and games, they next took possession of a small tent rigged up on the upper deck, to which the ejected smokers had retired. There are moments in such a voyage when one thinks that half a gale of wind with a cross sea would not be altogether unwelcome.

If such a perverse wish did arise in any breast, it was certainly disappointed. The voyage to Monte Video was uneventful, and offered little of special interest, but the weather was throughout fine. On the second day we met a slight breeze from the north, causing a decided rise of temperature and a fall of the barometer, but only a few drops of rain fell ; and then, after returning to the normal temperature, the thermometer rose steadily as we advanced daily about four degrees of latitude. It may be worth while to give the following extract from my notes, observing that on board ship temperature observations are merely rough approximations. Those best admitting comparison are made about a quarter of an hour after sunset, the precise hour, of course, varying with the latitude, of which I give only a rough estimate.

Date.	Time.	Latitude.	Barometer.	Thermometer (Fahr.).
June 16	Sunset	52° 30′	30·06 in.	37·5°
,, 17	Noon	50°	29·68 ,,	48·5°
	Sunset		29·70 ,,	48·0°
,, 18	Noon	46°	29·90 ,,	50·5°
	Sunset		29·90 ,,	45·2°
,, 19	9 a.m.		29·90 ,,	52·0°
	Noon	42°	29·86 ,,	
	Sunset		29·88 ,,	48·0°
,, 20	10 a.m.	38° 20′	29·88 ,,	
	Sunset		29·83 ,,	54·0°

Favoured by clear weather, we occasionally had glimpses of projecting headlands on the Patagonian coast, and especially on the 19th, when we made out the promontory of San José on the south side of the wide and deep Bay of San Matias, and later in the same day sighted some hills on the north side of the same gulf near the mouth of the Rio Colorado, the chief of Patagonian rivers.* As far as I could

* It is unfortunate that the Spaniards who had the naming of so large a part of the American continent should have shown so little inventive faculty. When they did not adopt a native name for a river, they rarely got beyond Red River, Black River, or Big River, and wherever we turn we encounter a Rio Colorado, a Rio Negro, or a Rio Grande.

discern, the sea-birds that approached the ship were the same species which had visited us on the Pacific coast, cape pigeons being as before the most numerous and persevering.

At sunrise on the shortest day we were approaching the city of Monte Video. Covering a hill some three hundred feet in height, and spreading along the shore at its base, the town presents a rather imposing aspect. It looks over the opening of the vast estuary of La Plata, fully sixty miles wide, into which the great rivers of the southern half of the continent discharge themselves. From the detritus borne down by these streams the vast plains that occupy the larger part of the Argentine territory have been formed in recent geological times, but the alluvial deposits have not yet filled up the gulf that receives the two great streams of the Paraná and the Uruguay. It would seem, however, that that consummation is rapidly approaching. Extensive banks, reaching nearly to the surface at low water, occupy large portions of the great estuary, and the navigable channel is so shallow that large ships are forced to anchor twelve or fourteen miles below Buenos Ayres, and even at Monte Video cannot approach nearer than two miles from the landing-place.

A small steam-tender came off to convey passengers to the city, and, with very little delay at the custom-house, I proceeded to the Hotel de la Paix, a French house, to which I was recommended. In spite of the irregularity of the ground, the city is laid out on the favourite Spanish chess-board plan, in *quadras* of nearly equal size. The main streets run parallel to the shore, and, being nearly level, are well supplied with tramcars ; but the cross streets are mostly steep and badly paved. The flat roofs of the houses, enjoying a wide sea-view, are the favourite resort of the inmates in fine weather, and many of them have a *mirador*, roofed in and windowed on all sides, whence idle people may enjoy the view sheltered from sun or rain. A stranger is at once struck by one marked difference between the towns on the Atlantic coast and those on the western side of South America. Here people live free from the constant dread of earthquakes, and do not shrink from making their town houses as high as may be convenient ; but the towns become more crowded, and one misses the charming *patios* of the better houses of Santiago and Lima.

To a traveller fresh from Peru and Chili and Western Patagonia, the region which I now entered, with its boundless spaces of plain and its huge rivers, appears by comparison tame and unattractive to the lover of nature. It is true that the industrial development of the last quarter of a century has been almost as rapid here as in the great republic of North America. The great plains are now traversed by numerous lines of railway, and steamers ply on the greater rivers and several of their tributaries. A naturalist may now accomplish in a few weeks, and at a trifling cost, expeditions that formerly demanded years of laborious travel. The southern slopes of the Bolivian Andes, stretching into the Argentine States of Salta, Oran, and Jujuy, are easily reached by the railway to Tucuman ; and yet easier is the journey by the Paraguay river steamers that carry him over seventeen hundred miles of waterway to Cuyabà, in Central Brazil, the chief town of the great province of Matto Grosso. But the time at my disposal was strictly limited, and the coming glories of Brazil haunted my imagination, so that I had no difficulty in deciding to make but a brief halt in this part of the continent, limiting myself to a short excursion on the river Uruguay and a glimpse of Buenos Ayres.

Of three days passed at Monte Video a considerable portion was occupied by the English newspapers, full of intelligence of deep and chiefly of painful interest ; but I twice had a pleasant walk in the country near the city. Some heavy rain had fallen before my arrival, and the roads, which are ill kept, were deep in mire ; but the winter season in this region is very agreeable, and the favourable impression made during my short stay was confirmed by the general testimony of the residents as to the salubrity of the climate. The winter temperature is about the same as in the same latitude on the Chilian coast, but the summers are warmer by 9° or 10° Fahr., and the mean temperature of the year fully 5° higher, being here about 62° Fahr. We are, however, far removed from the great contrasts of temperature that are found on the eastern side of North America. At Monte Video the difference between the means of the hottest and coldest months is 22°, while in the same latitude on the coast of North Carolina the difference is fully 35°. On the whole, the climate most nearly resembles that of places on the coast of Algeria, especially that of Oran, save that in the latter place the winters are slightly colder and the summer months somewhat hotter.

The town is surrounded by country houses belonging to the merchants and other residents, each with a *quinta* (garden or pleasure-ground), in which a variety of subtropical plants seem to thrive. Comparatively few of the indigenous plants showed flower or fruit, certainly less than one is used to see in winter nearer home on the shores of the Mediterranean. But a small proportion of the ground is under tillage, and beyond the zone of houses and gardens one soon reaches the open country, which extends through nearly all the territory of the republic. The English residents have adopted the Spanish term (*campo*), which is universally applied in this region of America to the open country whereon cattle are pastured, and the stranger does not at first well understand the question when asked whether he is "going to the camp."

The only fences used in a region where wood of every kind is scarce are posts about six feet high, connected by three or four strands of stout iron wire. These are set at distances of some miles apart, and serve to keep the cattle of each *estancia* from straying. It is said that when these fences were first introduced, many animals were killed or maimed by running at full speed against the iron wires, but that such cases have now become rare. The more intelligent or more cautious individuals avoided the danger, and have transmitted their qualities to a majority of their offspring.

At the hospitable table of the British minister, Mr. Monson, I met among other guests Mr. E——, one of the principal English merchants, whose kindness placed me under several obligations. On the following day he introduced me to an enterprising Italian, whose name deserves to be remembered in connection with modern exploration of the coasts of Patagonia and Tierra del Fuego. Signor Bartolomeo Bossi, who emigrated early in life to South America, seems to be a born explorer, and whenever he has laid by sufficient funds for the purpose he has forsaken other pursuits to start upon some expedition to new or little known parts of the continent. In a small steamer of 220 tons, fitted out at his own cost, he has in two expeditions minutely explored the intricate coasts of the Fuegian Archipelago and a great portion of the Channels of Patagonia.

Several of the discoveries interesting to navigators made in the course of the first of these voyages were published in the *Noticias Hidrograficas* of the Chilian naval department for 1876, and Signor Bossi asserts that the chief motive that determined the English admiralty in despatching the surveying expedition of the *Alert* was to verify the announcements first made by him. I have not seen any reference to Signor Bossi in the interesting volume, "The Cruise of the *Alert*," by Dr. Coppinger; but it appears certain that many of the observations recorded in the Santiago *Noticias* have been accepted, and are embodied in the most recent charts.

In this part of America the Republic of Uruguay is commonly designated as the Banda Oriental, because it lies altogether on the eastern bank of that great river. It possesses great natural advantages—

fine climate, sufficiently fertile soil, ready access by water to a vast region of the continent, along with a favourable position for intercourse with Europe. But these privileges are made almost valueless by human perversity. The military element, which has been allowed to dominate in the republic, is the constant source of social and political disorder. A stable administration is unknown, for each successful general who reaches the presidential chair must fail to satisfy all the greedy partisans who demand a share of the loaves and fishes. After a short time it becomes the turn of a rival, who, with loud promises of reform, and flights of patriotic rhetoric, raises the standard of revolt. If he can succeed in getting enough of the troops to join him, the revolution is made, and Uruguay has a new president, whose history will be a repetition of that of his predecessors. If the pretender should fail, he is summarily shot, unless he be fortunate enough to make his escape into the adjoining territories of Brazil or Argentaria.

On the day after my arrival the news of a rising headed by a popular colonel reached the capital, and troops were sent off in some haste to suppress the revolt. In each case the existence of the Government depends on the uncertain contingency whether the troops will remain faithful or will hearken to the fair promises of the new candidate for power.

It is obvious that a country in a chronic condition of disorder is a very inconvenient neighbour, and Uruguay would long have ceased to exist as a separate government, if it were not for the jealousy of the two powerful adjoining states. Brazil and

Argentaria* are each ready and willing to put down the *enfant terrible*, but neither would tolerate the annexation by its rival of such a desirable piece of territory. The prospect of a long and sanguinary war has hitherto withheld the Governments of Rio and Buenos Ayres, and secured, for a time, immunity to Uruguayan disorder.

I had arranged to start on the 24th of June, in the steamer which plies between Monte Video and the Lower Uruguay. That day being one of the many festas that protect men of business in South America from the risk of overwork, banks and offices were closed, and but for the kindness of Mr. E—— I should have found it difficult to carry out my plan. I went on board in the afternoon, and found a small crowded vessel, not promising much comfort to the passengers, but offering the additional prospect of safe guidance which every Briton finds on board a ship commanded by a fellow-countryman.

The sun set in a misty sky as we left our moorings and began to advance at half speed into the wide estuary of La Plata. As night fell the mist grew denser, and during the night and following morning we were immersed in a thick white fog. It was in reality a feat of seamanship that was accomplished by our captain. The great estuary of La Plata, gradually narrowing from about sixty miles opposite

* The constant inconvenience of employing such cumbrous expressions as Argentine Confederation or Argentine territory for a state of such vast extent and such yearly increasing importance must be felt by every one who has occasion to speak or write about this region of America. I trust that I shall be forgiven if in this book, as well as elsewhere, I have taken the liberty of applying a single name, which has nothing about it so strange as that it should not long since have come into use.

Monte Video to about sixteen at Buenos Ayres, is almost everywhere shallow and beset by sand or mud-banks, between which run the navigable channels. According to their draught, the ships that conduct the extensive trade between Buenos Ayres and Europe are spread over the space below the city, the larger being forced to anchor at a distance of fourteen miles. To avoid the banks, and to escape collision with the ships in the water-way, in the midst of a fog so dense was no easy matter. It is needless to say the utmost caution was observed. We crept on gently through the night, and at daybreak approached the anchorage of the large ships. Our captain seemed to be perfectly acquainted with the exact position of every one of them, and, as with increasing light he was able to recognize near objects, each in turn served as a buoy to mark out the true channel. Soon after sunrise we reached the moorings, about two miles from the landing-place, and lay there for a couple of hours, while the Buenos Ayres passengers and goods were conveyed to us in a steam-tender. It was a new experience to know one's self so close to a great and famous city without the possibility of distinguishing any object.

At about ten a.m. we were again under steam and making for the mouth of the Uruguay on the northern side of the great estuary. The fog began to clear, and finally disappeared when, a little before noon, we were about to enter the waters of the mighty stream, which is, after all, no more than a tributary of the still mightier Paraná.* Just at this point, signals and

---

* The Paraná, with its great tributary the Paraguay, drains an area

shouts from a very small steamer induced our captain to slacken speed. The strangers urgently appealed to him to take on board some cargo for a place on the river, the name of which escaped me. To this request a polite but very decided refusal was returned, the prudence of which we afterwards appreciated. The cargo in question doubtless consisted of arms, ammunition, or other stores for the use of the revolutionary force supposed to be gathered at Mercedes, not far from the junction of the Rio Negro with the Uruguay, and it clearly behoved the steamboat company to avoid being involved in such enterprises.

At its mouth the Uruguay has a width of several—probably seven or eight—miles, and at the confluence of the Rio Negro, some fifty miles up stream, the breadth must be nearly half as much. The water at this time was high, as heavy rain had fallen in the interior, and the current had a velocity of about three miles an hour. I believe that it is only exceptionally, during unusually dry seasons, that tidal water enters the channels of the Paraná or the Uruguay. I was struck by the frequent passage of large green masses of foliage that floated past as we ascended the river. Some consisted of entire trees or large boughs, but several others appeared to be formed altogether of masses of herbaceous vegetation twined together or adhering by the tangled roots. It can easily be imagined that, where portions of the bank have been undermined and fall into a stream, the soil is washed away from the roots, and the whole may be floated down the stream and even

of more than 1,100,000 square miles; the basin of the Uruguay is reckoned at 153,000 square miles.

carried out to sea. The efficacy of this mode of transport as one of the means for the dispersion of plants is now generally recognized, and, considering that the basin of the Paraná covers a space of over twenty-one degrees of latitude, we must admit the probability that it has had a large part in the diffusion of many tropical and subtropical species to the southern part of the continent.

The Rio Negro, which drains about half the territory of the republic, is the chief affluent of the Uruguay. At the junction we met a small steamer which plies to and fro on the tributary stream, and some time was lost in effecting the exchange of passengers and cargo. From some new-comers we gathered rather vague reports as to the attempted revolution. The chief was a certain Colonel Maximo Perez, already well known in Uruguayan political life. I have already explained that the term in this country means the effort to use the soldiery to upset the existing administration, or, if you happen to be in power, to employ the same agency to make short work of your rivals. It was generally thought that Perez had made the mistake of raising the standard too soon, and must fail. This anticipation was soon verified, and before I left the country two reports, each equally authentic, reached the capital—the one that he had made his escape, the other that he had been shot. To the community it was a matter of indifference which story might be true: in the one case, he would appear again to renew the revolt; in the other, some new adventurer would take his place.

A few miles above the confluence of the Rio Negro

we reached Fray Bentos, the great factory where "Liebig's Extract of Beef" is prepared and sent to Europe. Whatever prosperity exists in the Banda Oriental depends altogether on beef. To the raising of horned cattle the greater part of the soil of the republic is devoted, and in caring and guarding them most of the rural population is employed. The *saladeros*, where the animals are slaughtered and the various parts converted to human use, are the chief, almost the only, industrial establishments, and it is their produce that supports the trade and navigation.

Though the channel is narrower above the junction of the Rio Negro, the Uruguay was still a mighty river, from one to two miles in width, with numerous islands, all covered with trees and seemingly uninhabited. The trees on the islands and along the banks are mostly small, about thirty feet in height, but on some of the islands they must certainly surpass fifty feet. It was impossible for a passing stranger to identify the unfamiliar forms of these trees, which seemed to present considerable variety, the more so as the majority appeared to be deciduous, and but a few withered leaves remained on the nearly bare branches.

Paisandu, the place of my destination, is about a hundred and fifty miles from the mouth of the river, and the steamer often accomplishes the distance in fourteen hours. I was led to hope that we should arrive soon after midnight, but as night fell a dense fog spread over the river. Further progress was impossible, and we dropped anchor in mid-channel.

With sunrise the fog quickly melted away, and the turning of the screw soon announced that we had resumed our journey. Up to this point the banks of the river on either side had been absolutely flat, but at an early hour on the 26th we for the first time were relieved by the appearance of some rising ground on the east side of the river. There was nothing deserving to be called a hill, but so impatient is human nature of the monotony of dead-level, that even a rise of a couple of hundred feet is a welcome alleviation. A house on the summit, which must command a vast range of view, appeared to be the only desirable residence I had yet seen in this region. The dead-level soon resumed its place on the eastern bank; but a few miles farther we began to descry a range of low hills on the opposite, or Argentine, bank of the stream. We had hitherto held no communication with the territory on that side, but before noon we dropped anchor opposite to the landing-place for the town of Concepcion. This is one of the chief places in the state of Entrerios, which, as the name implies, fills the space between the two great rivers, Paranà and Uruguay, and extends northward about two hundred and forty miles from the estuary of La Plata. The town stands on a low hill about two miles from the river. Some passengers went ashore, a few were taken in their place, and after a short delay the screw was again in motion and the voyage was resumed.

About two p.m. we were at length opposite to Paisandu, a name known to most English readers only by the ox-tongues prepared at the neighbouring *saladeros*.

One of the peculiarities of this region arises from the fact that in the estuary and along the lower course of the great rivers the banks shelve so gradually that boats are seldom able to approach the shore. Elsewhere the inhabitants would make provision by constructing long jetties carried far enough to enable boats to draw alongside. But suitable timber is said to be scarce and very dear, and, besides, such constructions would deprive a part of the population of their means of gaining a livelihood. Carts with a pair of enormous wheels, seven or eight feet in diameter, are driven into the water till it reaches nearly to the shafts, and passengers scramble as best they may into or out of the boats. In this novel fashion I reached the shore, with one or two other passengers.

Paisandu has the aspect of a thriving country town, with streets and buildings of plain aspect, but looking clean and well cared for. It stands on rising ground, which is not a hill, but merely the river-ward slope of the flat country through which the Uruguay has here scooped a broad trench about a hundred feet below the general level. I found a very fair country inn kept by an Englishman, and at once proceeded to deliver a note of introduction to Dr. French, an English physician who enjoys considerable local reputation. The days being short at a season corresponding to our European Christmas, it was already too late for an excursion to the neighbouring country, which was postponed till the following morning; and I passed the greater part of the afternoon and evening in the agreeable society of Dr. French, whose range of general information, and thorough acquaintance with the country which he has made his home, rendered his conversation interesting and instructive.

Many Englishmen seem to imagine that, at least as regards material progress, distant countries, with the possible exception of the United States, are much less advanced than we are at home. I was led to an opposite conclusion as far as the more advanced states of South America are concerned, and I was struck by one illustration of the fact that I encountered at Paisandu. In the course of my long conversation with Dr. French, we were three times interrupted by the tinkling of a little bell connected with telephone wires carried into his sitting-room. I learned that a wire was carried from each of the chief *estancias* and *saladeros* within a circuit of eight or ten miles from the town. On each occasion advice was sought and obtained as to some case of sickness or accident, and it was impossible not to be struck by the great addition thus made to the usefulness of a skilful medical adviser in country districts. With regard to this and other applications of the telephone and the electric telegraph, our backward condition may be explained by the extraordinary fact that the English people have tolerated the existence of a Government monopoly, which, in many cases, acts as a prohibition; but in other matters, such as electric lighting, our relative inferiority must be set down to the extreme slowness with which new ideas germinate and reach maturity in the English nature.

I was much interested by the information given to me by Dr. French as to the frequent occurrence of the fossil remains of large extinct mammalia in this district. Complete skeletons are, of course, not commonly found; but large bones in good condition are, as I learned, easily procured. My stay was necessarily so short that I could not expect to obtain any, but I entertained a hope, not yet realized, that through the kind intervention of Dr. French, some valuable specimens might be obtained for the Cambridge University Museum. But to complete our knowledge of the very singular extinct fauna of this region of America, prolonged research on the spot, conducted by experienced palæontologists, is a necessary condition. These plains are the cemeteries in which myriads of extinct creatures lie entombed. We probably have got to know the majority of the larger species, but it is probable that many others have as yet escaped the notice of naturalists.

The steamer in which I had travelled ascends the river as far as Salto, about sixty miles above Paisandu; but at that place the navigation is interrupted by rapids, and travellers pursue their journey by land until they reach the steamers that ply on the upper waters of the Uruguay. I should have wished to visit Salto, but the steamer was to arrive at night and to depart on the return voyage next morning. By stopping at Paisandu I secured the opportunity for seeing a little of the country and the vegetation.

By way of seeing something of the natives, Dr. French took me to one of the best houses in the town, and introduced me to one of his patients, an old lady ninety years of age. She did much credit to the skill of her medical adviser, as I found her full of life and activity, conversing freely and intelligently on the topics of the day. In the garden surrounding her house were a number of orange trees in full bearing, and, amongst other exotics, the largest tree of *Eucalyptus globulus* that I have yet seen; though planted, as the old lady assured me, only twenty years before.

It was announced that the return steamer was due at two p.m. on June 27, so I arranged, in the language of this region, to go for an excursion *to the camp* as early as possible in the morning. In company with a young Englishman to whom Dr. French had introduced me, I started in a carriage, and, after passing through the belt of gardens and fields surrounding the town, soon reached a rather wide stream running between muddy banks. I now understood why all the vehicles here are hung upon such extremely high wheels. The horses take to the water as easily as if they were amphibious, and we got across the stream without taking in water, but not without a severe tug to get the carriage through the deep mud. We next approached a large *saladero*; but I had no curiosity to see the process of slaughter, nor the various stages by which a live animal is speedily converted into human food. We made a circuit round the *saladero* and the adjoining enclosures, and before long reached the open country.

The general aspect reminded me of what I have seen at the corresponding season in the less inhabited parts of Northern Africa, especially near Tunis, although the plants, as might be expected, are not only different, but in great part belong to different natural families. Open spaces covered with herbaceous vegetation alternate with patches of low bushes, mostly evergreen, and here and there with shrubs under ten feet in height; but there was nothing deserving to be called a tree. The indigenous trees of this region seem to be confined to the banks and islands of the great rivers. Among the bushes were four species of *Baccharis*, a Composite genus characteristic of South America, three species of *Solanum*, a *Lycium*, etc. But the commonest bush, which extends from the Tropic of Capricorn to Patagonia, is *Duvaua dependens*, with crooked branches beset with stout thorns, which has no near ally among European plants. I found several plants still in flower—two or three pretty species of the mallow tribe, a *Buddleia*, an *Oxalis*, and a *Verbena* (*V. phlogifolia*), nearly allied to the ornamental species of our gardens.

I returned to the town just in time to have all in readiness for the steamer, which arrived punctually at two o'clock, and, after bidding farewell to Dr. French, embarked with the impression that life in a country town on the Uruguay is very much like life in a country town anywhere in Europe—somewhat dull, but not devoid of interest to one who is content to feel that he has been of some use to fellow-creatures.

The weather had become brighter, and we were spared the annoyance of waiting at night for the clearing of the fog. We held on our course down the stream, and at sunrise were again at anchor opposite to the city of Buenos Ayres, now for the first time become visible. Seen in the bright morning light, it presented a somewhat imposing aspect, as befits the most populous and important port of the South American continent. The advance of the Argentine

Confederation has been so rapid since public tranquillity has been assured that the returns of a few years ago are doubtless considerably below the truth. Those of the five years from 1870 to 1874 show a yearly average of about ten millions sterling of imports, and nearly seven and a half millions of exports; but these figures, especially the latter, should now be much increased. Of the whole commercial movement more than eighty per cent. belongs to Buenos Ayres, and the extension of railways must further increase its supremacy.

I went to the Hotel de Provence, a French establishment fairly well kept, and, after confinement in the little den on board the river steamer, enjoyed the novel sense of occupying a spacious room. A good part of the day was spent in wandering about the town. It is built on the regular chess-board plan, with *quadras* of equal dimensions. The streets are narrow and ill-paved, most of them traversed by tramcars, which are the only convenient vehicles; but the whole place is pervaded by an air of activity which seems strange in Spanish America, reminding one rather of the towns of the United States.

I was directed to an exhibition of the natural products and manufactures of the states * of the Argentine Confederation, which appeared to make a creditable show, but of which I felt myself to be no competent judge. I was chiefly interested by the large collections of native woods from Corrientes and the mountain regions of Tucuman, Salta, and the adjoining states.

---

* The term *provinces*, commonly applied to the federated States, is misleading, and should be laid aside.

We know at present very little as to the extent of the Argentine forests, and still less as to the proportion in which the more valuable species are distributed; but it is obvious that in these forests there exist important sources of wealth, which, however, must require good management for their future development. Many of the largest and most valuable trees belong to the family of *Leguminosæ*, and may be found to rival in importance those of Guiana.

Speaking of the forests of the northern states, the late Professor Lorentz writes that they are exclusively confined to the eastern slopes of the mountains on which the winds from the Atlantic deposit their moisture, while the western slopes remain dry and bare of trees. He dwells on the need for an efficient forest law, as the result of the carelessness of the sparse population is that in the neighbourhood of inhabited places much valuable timber is ruthlessly destroyed. It may be feared that, under a constitution which, for such purposes, leaves practical autonomy to fourteen different states, it may be very difficult to obtain the enactment of an efficient law, and still more difficult to secure its enforcement.

The chief architectural boast of Buenos Ayres is the Plaza Mayor, one side of which is occupied by the cathedral, a very large pile in the modern Spanish style, which is not likely to serve as a model for imitation. The day being a *festa*, there was a ceremony in the afternoon, which attracted a crowd of the female population. The great church was ablaze with thousands—literally thousands—of wax candles, and the entire pavement was covered with costly carpets of the most gaudy colours. The behaviour of the congregation did not convey to a stranger the impression of religious feeling. It is doubtful, however, to what extent we are right in applying in such matters the standard derived from a different race and different modes of feeling. A severer style of worship would have no attractions for a people who thirst for satisfaction to the eye and ear ; and they would certainly not be the better, in their present condition of progress, if the scepticism of the age were to close this avenue of escape from the sordid cares of daily life.

On June 29, my second day at Buenos Ayres, I made a short excursion to the Boca, on the shore of the Rio de la Plata, only about three miles from the city. I had an illustration of the careless way in which, from want of sympathy or want of imagination, most people give directions to strangers. Being informed that the tramcars plying to La Boca were to be found in a certain street, I proceeded thither to look out for a vehicle going in the right direction. After a few minutes a vehicle appeared, coming from La Boca. After ten minutes more a second arrived from the same direction, and after ten minutes more a third, but not one in the opposite sense. At last I went into the shop of a German chemist near at hand, when the mystery was explained. The cars enter the town by one street, make a short circuit, and return by a different street.

The Boca does not offer much to interest a stranger. I could have fancied myself somewhere in the outskirts of Leghorn, so frequent were the familiar sounds of the Italian tongue, save that in Italy it would be difficult to find a spot where the horizon is unbroken by a near hill, or by the distant outline of Alp or Apennine.

Having paid a short visit to Mr. Schnyder, the newly appointed Professor of Botany, I strolled through the adjoining fields with the hope of finding some remains of the autumnal vegetation. The low flat country is intersected by broad ditches, and much reminded me of Battersea fields as they existed half a century ago, when I first began to collect British plants. Seeing in a ditch the remains of a fine *Sagittaria*, I filled a bit of paper with the minute seeds, and from these has sprung a plant which has for several seasons been admired by the visitors to Kew Gardens. It is the *Sagittaria Montevidensis*, which is not uncommon in Argentaria and Uruguay, but, so far as I know, does not extend to Brazil—a singular fact, considering that the seeds must be readily transported by water-birds. In its native home it grows to a somewhat larger size than the European species, but is not very conspicuous. Cultivated at Kew, in a house kept at the mean temperature of about 78° Fahr., it has attained gigantic proportions, rising to a height of over six feet, and the petioles of the leaves attaining the thickness of a man's arm.

I had arranged to take my passage to Brazil in the steamer *Neva*, of the Royal Mail Company, and at this season I felt no regret at quitting this region of South America, which offers comparatively slight attractions to the tourist. I was led, however, from all the information that I collected, to form a high estimate of the advantages that it offers to European settlers. At the present time the chief source of profit is from the rearing of cattle; but, though long neglected, agriculture promises to become the most important element of national prosperity. Until the middle of this century there were none but wooden ploughs of the type used by the aborigines, and corn was imported from abroad to feed the townspeople. There are now numerous agricultural colonies formed by foreign settlers, especially in the state of Santa Fé, and the results have been eminently successful. Large crops of grain, especially wheat, of excellent quality, are easily raised. The vine prospers, even as far south as Bahia Blanca, and in the northern states cotton, olives, tobacco, and other subtropical products appear to thrive. These agricultural colonies have been chiefly formed by Italian, Swiss, and German immigrants, and one of the most recent, composed of Welshmen, has been established so far south as the river Chubat in Patagonia. It may be feared that, owing to the deficient rainfall of that region, the prospects of the settlement are somewhat uncertain.

The Argentine Government has shown its wisdom in promoting immigration by the extraordinary liberality of the terms offered to agricultural settlers from Europe. With a territory as large as the whole of continental Europe, exclusive of Russia, and a population of scarcely two millions, immigration is the indispensable requisite for the development of resources that must render this one of the most important nations of the earth. The law, which, as I believe, is still in force, offers to settlers wishing to cultivate the national lands which are under the control of the Central Government the following terms:—An advance of the cost of the passage from a European port to Buenos Ayres, with conveyance from that city to the location selected; a free gift of a hundred hectares (about 247 acres) to each of the first hundred families proceeding to a new settlement; an advance, not exceeding a thousand dollars per family, to meet expenses for food, stock, and outfit, repayable without interest in five years; the sale of additional Government land at two dollars per hectare, payable in ten annual instalments; and, finally, exemption from taxes for ten years.

To the class of settlers who hold themselves above farming work other careers are open. Many young Englishmen who enjoy life in the saddle have done well as managers of *estancias*, for the raising of horses and cattle. The chief advice to be given to those who have some capital at their disposal is not to purchase property until they have gained practical experience. The Argentines show a laudable anxiety for the spread of education, and there is a considerable demand for teachers and professors, which has been mainly supplied from Germany, many of the professors from that country being men who have established a merited reputation.

One of the attractions of this region for European settlers is the excellence of the climate. Though not quite so uniform as that of Chili, it is free from the extremes of temperature that prevail in the United States. In the low country the difference between the mean temperature of the hottest and coldest months is from 22° to 25° Fahr., while in the middle states of the northern continent the difference is nearly twice as great—from 40° to 45°. The mean summer temperature is here about the same as in places six or eight degrees farther from the equator in eastern North America. The rainfall, which is of such vital importance to agriculture, appears not to be subject to such great annual irregularities as it is in the United States and Canada. The average at Buenos Ayres is about thirty-five inches annually, and in ascending the Paraná this increases to fifty-three inches in Corrientes, and eighty inches in Paraguay. It is only in some parts of the interior—*e.g.* about Mendoza—and in Patagonia, that the cultivator is, in ordinary seasons, exposed to suffer from drought.

Apart from the economic results of the great influx of immigration, the large recent admixture of European blood is effecting important salutary consequences. I have seen no recent returns, but it appears * that in the six years ending 1875, the number of immigrants from Europe exceeded 284,000, or about 47,500 annually; and I believe that this average has been exceeded since that date. Of the whole number fully one-half are Italians, and I found unanimous testimony to the fact that they form a valuable element in the population. With the exception of a small

---

* Much information respecting this country is to be found in a volume entitled, "The Argentine Republic," published in 1876 for the Centenary Exhibition at Philadelphia. It contains a series of papers prepared by Mr. Richard Napp, assisted by several German men of science.

proportion from the Neapolitan provinces, it is admitted that, whether as agricultural settlers or as artisans in the cities, the Italians are an orderly, industrious, and temperate class. The Germans and Swiss are not nearly so numerous, but form a useful addition to the orderly element in their adopted country. It may be hoped that experience and education have not been thrown away on the native Argentine, and that the memory of the forty years of intestine disorder which followed the final establishment of independence may serve as a warning against renewed attempts at revolution ; but assuredly the foreign element, which rapidly tends to become predominant, will be found an additional security against the renewal of disorder.

Although a majority of the large commercial houses at Buenos Ayres are English, and the trade with this country takes the first place in the statistical returns, the predominance is not so marked as it is on the western side of South America. Next to England, and not far behind, France has a large share in the trade, and although Germany has only lately entered the field, it appears that the business operations with that country are rapidly extending. Here, and at several other places in South America, I heard complaints that German traders palm off cheap inferior goods, having forged labels and trade-marks to imitate those of well-known English manufacturers. It is true that charges of a similar nature have been recently brought against some English houses. One asks if the progress of civilization is to lead us back to *caveat emptor* as the only rule of commercial ethics. If so, some further means must be discovered to enable the innocent purchaser to protect himself.

The most serious difficulty in the way of the increasing foreign trade of Argentaria is that arising from the shallowness of the great estuary of La Plata, which prevents large vessels from approaching the ports. In the course of ages nature will remedy the defect, when the present shoals are raised by deposits of fresh silt so as to confine the volume of water brought down by the great rivers, which would then scour out navigable channels. Whether the process may not be hastened by human skill and enterprise is a question which I am unable to answer. At present I believe that the only point where vessels of moderate burthen can approach the shore is at Ensenada, about fourteen miles below Buenos Ayres. It is now connected by railway with the capital, and promises to become an important trading port.

CHAPTER VI.

Voyage from Buenos Ayres to Santos—Tropical vegetation in Brazil—Visit to San Paulo—Journey from San Paulo to Rio Janeiro—Valley of the Parahyba do Sul—Ancient mountains of Brazil—Rio Janeiro—Visit to Petropolis—Falls of Itamariti—Struggle for existence in a tropical forest—The hermit of Petropolis—Morning view over the Bay of Rio—A gorgeous flowering shrub—Visit to Tijuca—Yellow fever in Brazil—A giant of the forest—Voyage to Bahia and Pernambuco—Equatorial rains—Fernando Noronha—St. Vincent in the Cape de Verde Islands—Trade winds of the North Atlantic—Lisbon—Return to England.

About midday on June 30, I took my departure from Buenos Ayres. The operation was not altogether simple or to be quickly accomplished. Jolting heavily over the ill-paved streets, a hackney coach carried me and a fellow-traveller with our luggage to the river-bank. The sight was very strange. It was a busy day, and there were literally hundreds of high-wheeled carts engaged in carrying passengers and goods out to the boats, which lay fully half a mile from the shore. When, after a delay that seemed excessive, we were installed in a boat, this was pulled in a leisurely fashion to the steam-tender, which lay more than a mile farther out. When the hour fixed for the departure of the tender was long past, we at length got under way, and finally reached the *Neva* steamship of the Royal Mail Company, about fourteen miles below the city, at five o'clock.

With iron punctuality dinner was served at the regular hour, although none of the passengers were ready, and the luggage was not brought on board till after dinner. There was, in truth, no reason for haste, as we were appointed to call at Monte Video on the following morning. My chief business at that place was to recover possession of the chest containing my botanical collections, which I had deposited at the custom-house.

Impressed with the attractions of Brazil, and feeling the strict limits of time to which I was bound, I asked myself if I should not have done better to have omitted a visit to the Plata region, and saved nine days by proceeding direct to Brazil in the *Iberia,* which started on the 22nd of June. I should certainly recommend that course to any naturalist travelling under similar circumstances at the same season ; but I am sure that, if I had done so, I should have felt regret at having missed an opportunity, and should have fancied that I had lost new and interesting experiences.

At four p.m. on the 1st of July the big ship began to move from her moorings opposite Monte Video, and for about sixty miles kept a due easterly course. Somewhere near the port of Maldonado we passed a bright light on an island which shows as a bold headland. I was told that this is known as Cape Frio, because of the cold often encountered here by those arriving from Brazil. It may be supposed that the force of the south-west wind which prevails in winter is more felt as the wide opening of the great estuary is reached. During my own short stay, the wind never rose beyond a gentle breeze, and the temperature on land was no more than agreeably cool, usually between 55° and 60° Fahr. during the day.

The distance from Monte Video to Santos, which is reckoned at 970 sea miles, was accomplished in about three days and eighteen hours. The voyage was uneventful. On the 3rd we approached the Brazilian coast, but the land lay low, and no objects could be distinguished. The weather was all that could be desired by the most delicate passengers, the barometer remaining almost stationary at about 30·2 inches,* and the temperature by day rising gradually from 57° at Monte Video to 62° in lat. 25° south. Before sunrise on the morning of July 5, we entered the bay through which the Santos river discharges itself into the Atlantic, and found ourselves in a new region. The richness of the green and the luxuriance of the foliage recalled the aspect of the coast at Jacmel, in Hayti, and as the morning advanced, while we slowly steamed towards the head of the bay, I had no difficulty in deciding on a course which had already suggested itself to my mind. I knew that Santos is connected by railway with São Paulo (better known in the form San Paulo), the chief town of this part of Brazil, and that the railway between that place and the capital was also completed; and I accordingly

* Dr. Hann ("Klimatologie," p. 657, *et seq.*) has discussed the causes of the prevalent high barometric pressure on both coasts of temperate South America, and has shown that in winter the area of maximum pressure moves northward towards the Tropic of Capricorn.

determined to leave the steamer, and find my way by land to Rio Janeiro.

Santos is an ancient place which had long remained obscure, until the great development of coffee-cultivation in South Brazil, and the construction of a railway to the interior, have made it the most advantageous port for the shipment to Europe of that important product. It lies at the mouth of an inconsiderable stream that enters the head of the bay. Seen from the sea, it appears to be backed by a range of lofty, flat-topped hills, but, in truth, these are no more than the seaward face of the great plateau which extends through a considerable part of the province of San Paulo. Although Santos is placed a few miles south of the Tropic of Capricorn, the aspect of the vegetation is completely tropical; and if a stranger were in doubt, the fringe of cocoa-nut palms on the shores of the bay would completely reassure him. Although the thermometer on board ship did not rise above 67°, the air seemed to us, arriving from the south, very warm, and we were surprised to hear the company's agent, when he came on board, complain that he had found the water in his bath uncomfortably chilly.

I landed with a young German fellow-traveller who, like myself, intended to proceed to San Paulo; and, as we found that the train was not to start for three hours, we occupied the time in ascending the nearest hill. It was now nearly three months since I had enjoyed a glimpse of true tropical vegetation in the forest of Buenaventura, and the interest and delight of this renewed experience can never be forgotten.

It was clear that on the slopes about Santos the native forests had been cleared, but on all the steeper parts, not reclaimed for cultivation, the indigenous vegetation had resumed the mastery. Trees and shrubs in wonderful variety contended for the mastery, and maintained, as they best could, a precarious struggle for existence with a crowd of climbers and parasites. So dense was the mass of vegetation that it was impossible to penetrate in any direction farther than a few yards, and there was no choice but to follow the track that led to the summit of the slope, on which stood a pretty house with an adjoining coffee-plantation. Among the many new forms of vegetation here seen, the most singular was that of the *Tillandsia*.*
Long, whitish, smooth cords hang from the branches of the taller trees, and at eight or ten feet from the ground abruptly produce a rosette of stiff leaves, like those of a miniature pineapple, with a central spike of flowers. But the most brilliant ornament of this season was a species of trumpet-flower (*Bignonia venusta*, Ker = *Pyrostegia ignea*, Presl), which, partly supporting itself, and partly climbing over the shrubs and small trees, covered them with dense masses of brilliant orange or flame-coloured flowers.

Laden with specimens, I returned to the town just in time for the afternoon train to San Paulo. The railway was constructed by an English company, and is so far remarkable that a somewhat difficult problem has been solved in an efficient and probably economical fashion. The object is, within a distance of a few

---

* The species common here is allied to *T. stricta*, but is not, I think, identical.

miles, to raise a railway train about 2500 feet. This is done by four stationary engines. The line is laid on four rather steep inclines, with nearly level intermediate spaces, each ascending train being counterpoised by one descending in the opposite direction, and the loss of time in effecting the connections is quite inconsiderable.

On every map of Brazil that I have seen, the Serra do Mar, which we were here ascending, is represented as a range of mountains running parallel to the coast, and extending from near Rio Janeiro to the Bay of Paranagua in South Brazil, apparently dividing the strip of coast from the low country of the interior. Most travellers would probably have expected, as I did, that on reaching the summit we should descend considerably before reaching San Paulo, and it was with surprise that from the summit I saw before me what appeared to be a vast level plain, with some distant hills or mountains in the dim horizon. It is true that the drainage of the whole tract is carried westward and ultimately reaches the Paraná; but the slope is quite insensible, and I do not think that, in the space of about sixty miles that lay between us and San Paulo, the descent can exceed two or three hundred feet. There was a complete change in the aspect of the vegetation, and open tracts of moorland recalled scenes of Northern Germany.

Night had closed before we reached the station at San Paulo. There was a difficulty about a carriage to convey us to the hotel. Perhaps the demands were unreasonable, or perhaps we were too unfamiliar with the coinage of Brazil, which is that of the mother country; but on hearing from the driver a demand for several thousand rees, we indignantly resolved to walk, and engaged a man to convey our luggage to the hotel. We were favourably impressed by the appearance of this provincial capital. In the space of a mile we passed through several good streets, well lighted with gas, and better paved than any I had seen in South America. Many handsome houses with adjoining gardens were passed on the way, and, on reaching the Grand Hotel, nice clean rooms, and good food provided for the evening meal, further conduced to favourable first impressions of Brazil.

My young German companion, a traveller for a commercial house, was returning from a visit to the interior of Brazil. By steamer on the Paraná and Paraguay he had gone from Buenos Ayres to Cuyabà, the capital of the province of Matto Grosso, a vast region with undefined boundaries, probably larger than most of the European states. I have often been struck by the results of superior education among Germans engaged in business, as compared with men of the same class in other countries. It is not that they often merit the designation of intellectual men, and still more rarely do they show active interest in scientific inquiry; but they retain a respect for the studies they have abandoned, are ready to talk intelligently on such subjects, and, as a rule, have a regard for accuracy as to facts which is so uncommon in the world, as much because the majority are too ignorant to appreciate their importance as owing to deliberate disregard of truth. I did not learn much as to the progress of inner Brazil, but my fellow-traveller mentioned a few particulars that had struck him as singular. He found the civil population of Cuyabà solicitous in their adherence to European fashions in dress, and, as a special note of respectability, the men always appearing in what are vulgarly called chimney-pot hats. The current coin in all but small transactions consisted in English sovereigns, but he was unable to explain how these have reached a region which can have so few commercial relations with this country. He departed on the following morning, while I resolved to spend a day in visiting the neighbourhood of the city.

Although San Paulo lies exactly on the southern tropic, the winter climate is positively cool, and at sunrise on July 6 the thermometer stood at 58° Fahr. On a rough estimate from a single barometric observation it stands about 2400 feet above the sea. Its appearance was altogether unlike that of all the towns seen in Spanish America. The somewhat wearisome monotony of regular square blocks gave place to the irregular arrangement of some of the provincial towns in England, several streets running out into the country and ending in detached villas. The general impression was that of comfort and prosperity. Several well-appointed private carriages were seen in the streets, and the shops were as good as one commonly sees in a European town of the same class.

I was much interested by the short country excursion, which occupied most of the day, and by an aspect of vegetation entirely new to me. The plants, with scarcely an exception, belonged to genera prevailing in tropical America, many of them now seen by me for the first time ; but the species were nearly all different from those of the coast region, and the general aspect of the flora still more markedly different. There was no trace of that luxuriance which we commonly expect in tropical vegetation ; monocotyledonous plants, except grasses, were very few, and, in place of the large ferns that abounded at Santos, I found but a single *Gleichenia*, allied to a species that I had gathered in the Straits of Magellan.

Although a fair number of plants were still in flower, I soon came to the conclusion that night frosts must be not unfrequent at this season, and that a considerable proportion of the vegetation must be annually renewed. I found several groups of small trees, chiefly of the laurel family, and for the first time saw the *Araucaria brasiliensis*, possibly in a wild state ; but none of the trees attained considerable height, and I doubt whether in a state of nature this plateau has ever been a forest region. I was rejoiced to see again, growing in some abundance, the splendid *Bignonia venusta*, and was led to doubt whether its real home may not be in the interior, and its appearance at Santos due to introduction by man.

We possess a fair amount of information as to the climate of the Brazilian coasts, but our knowledge of the meteorology of the interior provinces is miserably scanty. I was led to conjecture that, although the district surrounding San Paulo is not divided by a mountain range from the neighbouring coast region, the climate must be very much drier, and that the rainfall is mainly limited to the summer season.

In the course of my walk, I unexpectedly approached a country house about three miles from the town, and was somewhat surprised by meeting a carriage with ladies on their way to the house. As far as my experience has gone in the country parts of Portugal or Spain, such an encounter would there be regarded as a very unusual phenomenon.

The railway from San Paulo to Rio Janeiro appears to be a well-managed and prosperous concern, paying to its shareholders dividends of from ten to twelve per cent. The distance is about 380 miles, and the trains perform the distance in about thirteen and a half hours. Leaving my hotel in the dark, I found at the station a crowd of passengers contending for tickets; but good order was maintained, and we started punctually at six o'clock. For some way the line is carried at an apparent level over the plain, with occasional distant views of high hills to the north, and crosses two or three inconsiderable streams, whose waters run to the Paraná. A slight but continuous ascent, scarcely noticed by the passing traveller, leads to the watershed which, in this direction, limits the vast basin of the Paraná. After a long but very gentle descent, we reached a stream flowing westward. I at first supposed it, like those already seen, to be a tributary of the Paraná which made its way through some depression in the low ridge over which we had passed; but I soon ascertained that this was an error. Near the spot where the railway crosses it, the stream makes a sharp turn, and thenceforth proceeds in a direction little north of east for about four hundred miles, till it falls into the Atlantic at São João da Barra, north-east of Rio

Janeiro. This is the Rio Parahyba do Sul, not to be confounded with the Rio Parahyba north of Pernambuco, nor yet with the more important river Paranahyba in the province of Piauhy.

For the greater part of the way to Rio the railway runs parallel with the river. As laid down on the maps, the valley lies between a mountain range called the Serra da Mantiqueira on the left, and a minor range, which divides the upper course of the river from the middle part, where it flows in the opposite direction. The appearance of the country through which I now passed forcibly suggested to me views respecting its geological history, which were confirmed and extended by what I afterwards saw in the neighbourhood of Rio, and by all that I have since been able to learn on the subject.

I had already been struck by what little I had seen of the plateau region of the province of San Paulo. Beneath the superficial crust of vegetable soil, the plateau appears to be formed of more or less red arenaceous deposits, such as would result from the erosion and decomposition of the gneiss or granite which is the only rock I had seen in the country. In the valley of the Parahyba, the connection was unmistakable. Every section in the valley showed thick beds of the same coarse-grained, red arenaceous deposits, and on the slopes the same material lay at the base of whatever masses of granite we approached. But what especially struck me were the forms and appearance of the mountains on either hand, if that designation could properly be given to them. I saw nothing that would elsewhere be called a mountain range. The outlines were in most places rounded and covered with vegetation, but at intervals occurred steep conical masses, of the same general type as the sugar-loaf peaks surrounding the Bay of Rio Janeiro. However steep, the rocks nowhere showed angular peaks or edges, these being always more or less rounded.

It would be rash to generalize from the partial observations of a passing traveller; but the broad outlines of the geology of Brazil, or, at least, of the eastern provinces, have now been well traced,* and some general conclusions may safely be drawn. It is true that large districts of the interior have been but partially explored, and remain blanks on the geological map; but the eastern half of Brazil is undoubtedly ancient land, presenting no trace of secondary strata except in small detached areas near the coast, and where more recent tertiary deposits are to be found only in a portion of the great valley of the Amazons. A mountain range, having various local designations, but which may best be called the Serra da Mantiqueira, extends from the neighbourhood of San Paulo to the lower course of the Rio San Francisco, for a distance of twelve hundred miles, and this is mainly composed of gneiss, sometimes passing into true granite, syenite, or mica schist; and the same may be said of the Serra do Mar, a less considerable range lying between the main chain and the coast. The

* The best general account of the geology of Brazil that I have seen is contained in a short paper by Orville A. Derby, entitled, "Physical Geography and Geology of Brazil." It was published in the *Rio News*, in December, 1884, and, through the kindness of Mr. Geikie, I have seen a reprint in the library of the School of Mines.

southern limits of the Serra do Mar do not appear to be well-defined, but we may estimate its length at from five to six hundred miles. The other mountain systems of the empire are less well known; but I believe that the ranges dividing the province of Minas Geraes from Goyaz, and the so-called Cordillera Grande of the province of Goyaz, lying between the two main branches of the great river Tocantins, are largely formed of ancient sedimentary rocks of the Laurentian and Huronian groups.

The granite of the Serra da Mantiqueira and Serra do Mar is coarse-grained, with large crystals of felspar, and is therefore much exposed to disintegration. So far as I know, the vast masses of detritus forming the plateaux of this region show no other materials than such as would be produced by the disintegration of the crystalline rocks, and there is strong reason to believe that these have never been overlaid by sedimentary deposits.

Let us now consider what must have been the past history of a region formed of such materials, exposed, during a large part of the past history of the earth, to the action of the elements. In such an inquiry one of the chief points for consideration is the amount of rainfall. The direct effect, both mechanical and chemical, of rain falling on a rock surface is perhaps not the most important. Still more essential is its action in removing the disintegrated matter, and thereby exposing a fresh surface to renewed action. The difference in the absolute result due to abundant or deficient rainfall would be found, if we could calculate it accurately, to be enormous. In a nearly rainless country, such as Egypt or Peru, we see a slope covered with *débris*, and are apt to conclude that the rock is being rapidly disintegrated; but, in truth, what we see is the work of many, perhaps many hundred, centuries, which remains *in situ* because there is no agency to remove it. In a land of heavy rainfall the *débris* is speedily carried to lower levels, and the work of destruction is constantly renewed.

We have scarcely any observations of rainfall in the mountain districts of Brazil. The only reliable return that I have seen is that of one year's rainfall at Gongo Seco, in Goyaz, which amounted to more than a hundred and thirty inches; but we may safely conclude that it is everywhere very great. It is also important to note that if, as most geologists now believe, the Atlantic valley has existed since an early period of the earth's history, Eastern Brazil must always have been a land of heavy rainfall. A great mountain range on the eastern side of the continent might have created a desert region in the interior, but would have received in the past as much aqueous precipitation as it does at the present time.

We have, therefore, to consider what must have been the ancient condition of a region subjected throughout vast periods of geological time to the utmost force of disintegrating agencies applied to a rock very liable to yield to them, and where, without reckoning the large proportion which must have been carried by rivers to the sea, we see such vast deposits of the disintegrated materials formed out of the same matrix. To my mind the conclusion is irresistible that ancient Brazil was one of the greatest mountain regions of the earth, and that its summits may very probably have exceeded in height any now existing in the world. What we now behold are the ruins of the ancient mountains, and the singular conical peaks are, as Liais has explained, the remains of some harder masses of metamorphic gneiss, of which the strata were tilted at a high angle. As the same writer has remarked, although the crystalline rocks are for the most part easily disintegrated, some portions are formed of much more resisting materials, and these have to some extent survived the incessant action of destructive forces.

We are far from possessing the materials for a rational estimate of the probable extent and elevation of the ancient mountain ranges of Brazil. In the first place, we have a plateau region occupying a large part of the upper basin of the Paraná, with an area of fully 100,000 square miles, covered with detritus to an unknown, but certainly considerable, depth. In addition to this, it cannot be doubted that the finer constituents carried down by that river, and its tributary the Paraguay, from the same original home, have largely contributed to the formation of the Argentine pampas and Paraguay, including the northern portion of the Gran Chaco. Borings and chemical analysis of the soil may hereafter give us reliable data; but in the mean time we may safely reckon that an area of 200,000 square miles has been mainly formed from the materials derived from the ancient mountains whose importance I endeavour to point out. In addition to all this, we should further reckon the soluble matter and fine silt carried to the ocean during the long course of geological history, and take into account that the same great mountain region also furnished materials to streams which flowed northward and eastward.

In attempting to speculate on the past history of this region it is important to remark that, so far as evidence is available, there is reason to believe that Brazil has undergone less considerable changes of level than most other parts of the earth's surface. Even if we go back to the period of the earlier secondary rocks, there is no evidence to show that movements of elevation or depression have exceeded a few hundred feet.

I have attempted elsewhere[*] to give a sketch of the views which I hold as to the probable origin of the chief types of phanerogamous vegetation. I there pointed out that, at a period when physical conditions in the lower regions of the earth's surface were widely different, and the proportion of carbonic acid gas present in the atmosphere was very much greater than it has been since the deposition of the coal measures, it was only in the higher region of great mountain countries that conditions prevailed at all similar to those now existing. I further argued that, if the early types of flowering plants were confined, as I believe they were, to the high mountains, we could not expect to find their remains in deposits formed in shallow lakes and estuaries until after the probably long period during which they were gradually modified to adapt them to altered physical conditions.

[*] *Proceedings of the Royal Geographical Society* for 1879, p. 564.

A general survey of the South American flora shows, along with elements derived from distant regions, a large number of types either absolutely peculiar to that continent, or which, in some cases, appear to have spread from that centre to other areas. Of these peculiar types some may probably have originated in the Andean chain, but as to the majority, it seems far more probable that their primitive home was in Brazil; and it is precisely on the ancient mountains of this region that I should look for the ancestors of many forms of vegetation which have stamped their character on the vegetation of the continent.

I should be the first to admit that the views here expressed have no claim to rank as more than probable conjectures; but I hold that these, when resting on some positive basis of facts, are often serviceable to the progress of science, by stimulating inquiry and leading observers to co-ordinate facts whose connection had not previously been apparent.

In following the valley, in places where the siliceous soil supported only a scanty vegetation, I was struck by the singular appearance of scattered piles, usually about four feet in height, having much the appearance of rude milestones, occurring here and there in some abundance, but never very near each other. I was often able to avail myself of the short halts of the train at wayside stations to secure specimens of interesting plants, but I was not able to approach near to these unknown objects. I have no doubt, however, that they were habitations of termites, or, as they are commonly called, white ants. I have never been able to conjecture the origin of the instinct that induces so many species of termites in different parts of the world to construct dwellings in this form, nor what advantage they can derive from it.

As the Parahyba appeared to be a rapid-flowing stream, it is probable that in following the valley the railway descends considerably before it reaches the point, about eighty miles north of Rio, where it abruptly turns away from the river to make its way to the capital. The appearance of the vegetation announced a change of climate, but I did not notice any palms by the way. The country between the Parahyba valley and the coast appears to be an irregular mountain tract, nowhere of any great height, with projecting summits rising here and there of the same general character as those already described, and the railway follows a sinuous course so as to select the lowest depressions between the neighbouring bosses of granite. As we wound to and fro, constantly changing our direction amid scenes of increasing loveliness, night closed with that suddenness to which one becomes accustomed in the tropics, and the last part of the way was unfortunately passed in darkness. The approach to Rio must be surpassingly beautiful, but, beyond the fantastic outlines of the surrounding mountains, little could be discerned save the lights of the city, visible for many miles before we reached the railway station.

After a long drive through paved streets, I reached the English hotel (Carson's), and was curtly informed that the house was full. The next in rank is the Fonda dos Estrangeiros, to which I proceeded, and found quarters in a rather shabby room, not over-clean. The general style of the establishment and the food provided answered the same description. It is generally admitted that the accommodation for strangers in the capital of Brazil does not come up to the reasonable expectations of travellers.

By quitting the steamer at Santos, and travelling to Rio by land, I had gained some slight acquaintance with a new region, but I was well aware that I had suffered a considerable loss. The view on first entering the Bay of Rio Janeiro is one of those spectacles that leave an ineffaceable impression even on persons not very sensitive to natural beauty, and one on which my fancy from early youth onwards had most often dwelt. The pursuits of a naturalist, besides their own fascination, offer additional rewards to all who worship in the temple of Nature, but they also sometimes exact a sacrifice. Sallying forth on the morning of July 8, a little under the impression of the unattractive quarters of the night, I had but very moderate expectations as to what might be enjoyed of the scenery in the midst of a large city and its surroundings, but I was speedily disabused. Man has certainly done little to set off the unequalled fascinations of the place, but he has been powerless to conceal them. I passed a delightful day, partly strolling much at random on foot, and occasionally availing myself of the street-cars, which are frequented by all classes, and afford a stranger the best opportunity for seeing something of the very mixed population.

The famous Bay of Rio Janeiro may properly be described as a salt-water lake, so completely is it landlocked and cut off from the open sea. About thirty miles long and twenty in breadth, it is large enough to allow of spacious views, yet not so large as to lose in distance the marvellous background that is presented in every direction by the fantastic peaks that surround it. Numerous islands stud the surface, the larger telling their history in piles of huge blocks, either simulating rude Cyclopean architecture, or lying in wild confusion—granite pinnacles, half-decayed or fallen into utter ruin. The shores are everywhere a maze of coves and inlets, in which land and water are interlaced; and over all—the mainland and the islands alike—the wild riot of tropical vegetation holds its sway, defying the efforts of man to tame it to trimness. Even within the limits of the city, which stretches for about four miles along the shore, four or five coves present a ceaseless variety of outline. Of necessity the plan is completely irregular. Where a space of level ground opens out between the shore and the rocks, the city has spread out; where the rocks approach the water's edge, it is narrowed in places to a single street. In architecture, since the great era of Alcobaça and Batalha, the Portuguese have not achieved much, and their descendants in South America have done little to adorn the capital of their great empire. The largest building, the imperial palace, might easily be taken for a barrack. Nature has undertaken the decoration of the city, and, amid the palms, and under the shade of large-leaved tropical trees in the public walks and gardens, the absence of sightly buildings is not felt.

The suburb of Botafogo, which is the fashionable

quarter, lies on the shores of the most beautiful of the coves round which the city has grown up. It mainly consists of a range of handsome villas facing the sea, each with a charming garden, and, in this season, must be a delightful residence. But it is generally admitted that the climate of Rio is debilitating to European constitutions. As compared with most coast stations in the tropics the heat is not excessive—the mean temperature of the warmest month (February) is not quite 80° Fahr., and that of the coldest (July) about 70°; but most Europeans, and especially those of Germanic stock, require to be braced by intervals of cold, if they are to endure a hot climate with impunity. The annual appearance of yellow fever in the city supplies a still stronger motive to many of the foreign residents for fixing their abode amongst the hills. The chief resort, which in summer is frequented by most of the wealthier classes, is the well-known Petropolis, in the Organ Mountains, or Serra dos Orgãos, that rise beyond the northern shores of the bay.

From Botafogo I directed my steps towards the Botanic Garden, and, as usual among people of Portuguese descent, found great readiness in giving information to strangers. Following a road that turned away from the shore, I seemed to have left the city far behind, and be quite in the country; but presently another beautiful dark blue cove opened out before me, and again turning inland I reached the garden. I must confess to a feeling of something like disappointment at the famous avenue of palms. It has been correctly described as reproducing the effect of the aisle of a great Gothic cathedral, and the defect, as it seemed to me, is that the reproduction is too faithful. The trees of *Oreodoxa regia*, which are about a hundred feet in height, are all exactly of the same form and dimensions, so much alike that they appear to have been cast in the same mould, and it is difficult to persuade one's self that they are not artificial productions. It may not be easy to say why the same uniformity which satisfies the eye in a construction of stone, should fail to do so when similar forms are represented by natural objects. I suppose the fact to be that in all æsthetic judgments the mind is unconsciously influenced by trains of association. Our admiration is aroused not merely by given combinations of colour or form—by the mere visual image formed on the retina—but is controlled by our sense of fitness. We should resent as a caprice of the architect an irregularity in a vista of arches: among objects endowed with life we expect some manifestation of the universal tendency to variation.

With an intention, never fulfilled, to make a second visit to the garden, and, under the guidance of the director, Dr. Glaziou, to make nearer acquaintance with some of the vegetable wonders there brought together, I returned to my hotel. Before reaching Rio, I had decided to devote most of my short remaining time to a visit to the Organ Mountains, and to make Petropolis my head-quarters. As there was no especial reason for delay, I started for that place on the morning of the following day, July 9.

I shall make no attempt to describe the beauties of the bay as they were successively unfolded during the short passage to and from Petropolis. From early youth the Bay of Naples has ever appeared to me so perfectly beautiful that I was very reluctant to admit the pretensions of a rival. Even now I can well understand that some may find the pictures presented to the eye on the charmed coasts of our Mediterranean bay more complete, and the tints of the shores and sea and sky more harmonious; but there could be no doubt as to the gorgeous vesture that everywhere adorns this land. The vegetation of the Mediterranean coasts seems but poor and homely after the eye has dwelt on the luxuriance of tropical life, as though one were to compare a garb of homespun with trappings of velvet and embroidery. The islands of the bay present a ceaseless variety. Some are mere rocks, on which sea-birds of unknown aspect stood perched. Many of the larger are inhabited, and one, as I heard, has a population of thirteen hundred souls, and several charming villas showed it to be a favourite resort.

In about an hour and a half from the city, the little steamer ran alongside of a wooden jetty at a spot on the northern side of the bay facing the bold range of the Organ Mountains, which extend for over twenty miles in an easterly direction. Between the northern shore and the foot of the mountains is a level swampy tract, evidently filled up by the detritus borne down by the numerous streams, and beyond this the mountain range rises very abruptly from the plain. Somewhat to my disappointment, I ascertained that Petropolis lies at a considerable distance from the higher part of the Organ range to which my attention had hitherto been directed. It is towards its eastern extremity that the Serra shows that remarkable series of granitic pinnacles of nearly equal height, appearing vertical from a distance, that suggested the likeness to the pipes of an organ whence these mountains obtained their name. The height of the loftier part has been estimated at 7500 feet above sea-level. I do not think that any of the summits near Petropolis can surpass the level of 5000 feet.

A short train with a small locomotive carried passengers for Petropolis across the low tract to the point where the ascent abruptly commences, a distance of nine or ten miles. The marshy plain is doubtless fever-stricken, and we passed very few houses on the way to the terminus, which is appropriately named Raiz da Serra. The construction of a railway on the slope leading thence to Petropolis, up which trains should be drawn by a wire rope, had been commenced, but at the time of my visit passengers were conveyed in carriages, each drawn by six or eight mules. A well-kept and well-engineered road—by far the best mountain road that I have seen in any part of America —leads to the pass or summit of the ridge that divides Petropolis from the Bay of Rio. The views during the ascent, especially in looking back over the bay, were entrancing, and new and strange forms of vegetation showed themselves at each turn of the road. From the summit, a gentle descent of a couple of miles leads to the main street of Petropolis.

The place lies about 2900 feet above the sea, in a basin or depression amidst forest-covered hills.

The abundant rains of this region have carved the surface into a multitude of little dells and recesses, separated by hills and knolls of various size and height, leaving in their midst one comparatively broad space, where most of the buildings are grouped. The streamlets that issue from every nook in the mountains are finally united in two streams that flow in opposite directions, but both, I believe, ultimately find their way northward to the Parahyba. The streamlets have been turned to account by the inhabitants, for on each side of the main streets a rivulet of crystal water serves to maintain the vigour of a line of trees supplying the one need of the long summer—shelter from the vertical midday sun. In the present season (mid-winter) only one hotel was open; but in summer, when all who can do so escape from the oppressive heat of Rio, two or three others are generally crowded. It is at once apparent that Petropolis is a place for rest and enjoyment, not for business. The few shops and hotels are all in the main street, Rua do Imperador; the other streets, or roads, lie between ranges of detached villas, each with a garden, and here and there some more secluded habitation is withdrawn into some nook on the margin of the forest.

The large majority of the trees and shrubs of this region have persistent leaves, but a few lose their foliage annually in winter, and a few others, I believe, during the heat of summer. The only prominent reminder of the fact that we were in winter was the appearance of the *Bombax* trees that line the main street, now completely bare of foliage. The tree commonly planted in this part of Brazil is, I believe, the *Bombax pubescens* of botanists. The fruit, with its copious silky appendage to the seeds, alone remained at this season; but when covered with a mass of large white flowers, it must have a gorgeous appearance.

I cannot feel sure that every naturalist will approve of the resolution, which I very soon formed, to remain as long as was possible at Petropolis. To reach the higher summits of the Organ Mountains would have required at least three or four days' travel, and at this season I could expect to see very little of the vegetation of the higher zone. In the mean time, I found in the immediate neighbourhood, within a radius of four or five miles, an unexhausted variety of objects of interest, and the attractions of the place were doubtless heightened by the fortunate circumstances in which I found myself. It is certain that the ten days that I spent at this fascinating spot remain in my memory as the nearest approach to a visit to the terrestrial paradise that I can expect to realize. Besides the British minister, Mr. Corbett, I was fortunate enough to make the acquaintance of two English families, whose constant kindness and hospitality largely contributed to the enjoyment of my stay. To find in the midst of the marvels of tropical nature the charms of cultivated society, was a combination that I had not ventured to promise to myself.

Although I never went farther than five or six miles from my head-quarters, the variety of delightful walks in every direction seemed to be inexhaustible; go where one would, it seemed certain that one could not go wrong. I soon ascertained, indeed, that it is useless to attempt to penetrate the forests, except by following a road or cleared path. My first lesson was on the slope of a little hill some three hundred feet in height that overlooks the town. I was told that there was a path on the farther side, but, seeing the ground partly open, with trees of small stature not much crowded together, I resolved to follow the straight course. The ascent cost me over two hours of hard work, and I accomplished it only with the help of a sharp knife, by which to cut through the tangle of vegetation. In the midst of this I was surprised to find tall fronds of our common English bracken (*Pteris aquilina*), a fern that has been able to adapt its constitution to all but the most extreme climates of the world. The little hill that cost me so much labour had been completely cleared ten years before, so that all the trees and shrubs had grown up since that time.

The first excursion recommended to every stranger at Petropolis is that to the Falls of Itamariti. I went there twice, varying somewhat my course—the first time with a horse, which I found quite unnecessary and rather an incumbrance; the second time alone. The falls are not very considerable. A stream so slender that it can be passed by stepping-stones falls over two ledges of granite rock, together about forty feet in height; but, framed in a mass of the most luxuriant tropical vegetation, the whole forms a lovely picture. For some reason which I did not learn, the forest on the slopes of the lower part of the glen below the falls had been felled just before my visit, and its beauty had vanished, but fortunately the arm of the destroyer was arrested before reaching the falls

As happens to every stranger in a tropical forest, I was bewildered amidst the great variety of trees that struggle for supremacy, the one condition for victory being to get a full share of the glorious sunshine overhead. By vigorous tugging at one of the *lianes* that hung like a rope from a branch sixty feet above my head, I succeeded in breaking off a fragment, and identifying one of the larger trees as a species of fig, with large, oval, leathery leaves somewhat like those of a magnolia. It is needless to say that each tree is invaded by a host of enemies—parasites that fatten on its substance, comparatively harmless epiphytes that cling to the branches, and hosts of climbing lianes that mount to the topmost branches, robbing them of their share of sunlight, and hang down, often twined together, and in the deep shade are generally mere bare flexible stems. It was strange to observe that one of the deadliest enemies, a small parasite, fixing itself near the ground on the trunks of the larger trees, is a species of fig, belonging to the same genus as some of the giants of the forest, and doubtless tracing its descent from a common ancestor. It is in the tropical forest that one feels the force of Darwin's phrase "struggle for existence," as applied to the vegetable world. In our latitudes it is by an effort of the imagination that we realize the fact that in our fields and woodlands there is a contest going on between rival claimants for the necessary conditions of life. Here we see ourselves in the midst of a scene of savage warfare. The great climbers, like monstrous boas, that twine round and strangle the branch, remind one of the Laocoon; the obscure parasite that eats into the trunk of a mighty tree till a great cavity prepares its downfall, testifies to the destructive power of an insidious enemy.

It is only in the more open spots that a botanist is able to make close acquaintance with the smaller trees and shrubs. Near to the stream I was able to hook down a branch and secure flowering specimens of a *Begonia* that grew to a height of over twenty feet. In such situations *Melastomaceæ* were everywhere abundant, but for variety of forms the ferns surpassed any of the families of flowering plants. I was surprised to find that the beautiful tree ferns, that add so much to the charm of the tropical flora, were rarely to be found with fructification, and the huge fronds being of quite unmanageable dimensions, I did not attempt to collect specimens. Of the smaller kinds, when I was able, with the kind assistance of Mr. Baker, of Kew, to name my specimens, I found that I had collected thirty-five species in the neighbourhood of Petropolis.

During my stay here I visited a German gentleman whose singular manner of life excites the interest and curiosity of the European residents. I am ignorant of the motives that have led Mr. Doer, evidently an educated and cultivated man, to lead the life of a hermit far from his native country. He has built for himself a small house in the forest, on one of the hills that enclose the basin of Petropolis, and lives quite alone, except for the daily visit of a boy who carries the provisions that satisfy his very moderate wants. He seems to be entirely occupied in studying the habits of the native animals of the country, and especially those of the Lepidopterous insects—butterflies and moths—that adorn this region. By attention to the habitual food of the various species, he has succeeded in keeping in his house the caterpillars that in due time produce the perfect insect, and has preserved in cabinets large collections of fine specimens.

At the suggestion of the friend who accompanied me, Mr. Doer was good enough to introduce me to the family of small monkeys which he has raised and domesticated. The senior members had been brought from some place in Northern Brazil, but they had multiplied, some of the offspring being born in his house, and now formed a rather numerous party. The creatures habitually passed the day in the forest, never, in Mr. Doer's belief, wandering to a distance from the house, and at night came in and nestled among the rafters of the roof. The call was by a peculiar note, somewhat resembling a low whistle, repeated two or three times, and before a minute had elapsed the little creatures came swarming about the open window. They were decidedly pretty, their large black eyes giving an impression of intelligence, but I did not detect any indication of attachment to their master. I cannot say to what species they belonged. They had large ears like those of the marmoset, but differed in having a prehensile tail. One of them hung with his head downward, suspended by the tail from some projection above the window. After receiving some fragments of sweetmeat they soon departed, returning to their favourite haunts among the trees of the forest.

Starting early one morning, and reaching the crest of the range that divides Petropolis from the Bay of

Rio Janeiro, I enjoyed in great perfection a spectacle that is commonly visible at this season when the weather is clear and settled. Before sunrise a stratum of mist extends over the bay and the low country surrounding it. As I saw it, this may have been about a thousand feet in thickness when the sun first reached it, and the fantastic summits of the mountains rose like islets from a sea of dazzling white. As the sun's rays began to act, the mist appeared to melt away from above; the lower hills and the rocky islands of the bay emerged in succession, and finally the veil completely disappeared, and the whole wondrous view was completely disclosed.

The beautiful effects displayed in the gradual disappearance of mist as seen from a height in early morning must be familiar to every genuine mountaineer, and may be enjoyed amongst the hills of the British Islands. Among my own recollections, a certain morning, when I stood alone at sunrise on the highest peak of the Pilatus, near Lucerne, showed the phenomenon in a most striking way, accompanied as it was by the coloured halo that surrounds the shadow of the observer thrown on the cloud-stratum below. But in my previous experience the disappearance of the mist was always accompanied by the upward movement of some portions of the mass. The surface appears to heave under the action of force acting from below, and some masses are generally carried up so as temporarily to envelope the observer. In the view over the Bay of Rio I was much farther away from the surface of the mist than in previous experiences of the kind, and I may have been misled by distance from the scene of action, but, though watching attentively, I saw no appearance of heaving of the surface or any break in its regular form. The waste seemed to proceed altogether from the upper surface, and the emergence of the prominent objects in regular succession gave direct evidence to that effect.

During the first five days of my visit the weather at Petropolis was perfectly enjoyable. The temperature varied from about 60° Fahr. at sunrise to about 70° in the afternoon; but the effect of radiation must have been intense, as in an exposed situation a minimum thermometer descended on one night to 46°, and on the next to 44°, and the dew was heavier than I have ever seen it elsewhere, so that in some places the quantity fallen from the leaves of the trees made the ground perfectly wet in the morning. The barometer varied very little, even after the weather changed, and stood as nearly as possible three inches lower than at Rio, showing a difference of level of about 2900 feet. On the 16th of July the sky became overcast, and some rain fell in the afternoon, the thermometer rising at two p.m. to 73° Fahr., and moderate rain fell on each succeeding day until the evening of the 19th, but scarcely any movement of the air was perceptible. There is a remarkable difference in the distribution of rainfall between the part of Brazil lying within about fifteen degrees of the equator and the region south of that limit. At Pernambuco (south lat. 8° 4'), out of an annual rainfall of about a hundred and ten inches, nearly ninety inches fall during the six months from March to August, and at Bahia, with less total rainfall, the proportion is nearly the same.

But at Rio Janeiro the rainy season falls in summer, from November to March, and winter is the dry season. Of an annual rainfall of forty-eight and a half inches, only five and a half inches fall in the winter months, June, July, and August, and less than an inch and a half in July. No doubt the amount of rain is greater at a mountain station such as Petropolis, while the proportion falling in the different seasons must be about the same.

At Petropolis, as well as elsewhere in South America, I was struck by the fact that the children of European parents born in the country speedily acquire the indolent habits of the native population of Spanish or Portuguese origin. The direct influence of climate is doubtless one cause of the change of disposition, but I suspect that the chief share is due to the great difference in the conditions of life which are the indirect results of climate. Where mere existence is so enjoyable, where physical wants are so few and so easily supplied, the chief stimulus to exertion is wanting, and the natural distaste for labour prevails over the hope of gain. A boy will prefer to pick up a few pence by collecting flowers, or roots, or butterflies in the forest near his home, to earning ten times as much by walking to a distance, especially if expected to carry a light weight. On my first visit to Itamariti I took with me a German boy, whom I left in charge of the superfluous horse that I had been advised to take with me. Finding the occupation a bore, and probably fearing that he would have to carry back the portfolio and vasculum that I had taken for plant-collecting, he fastened the bridle to a tree and disappeared, never coming to claim the pay promised for his unaccomplished day's work.

All delightful times come to an end, and, as I resolved to visit Tijuca before departing from Brazil, I quitted Petropolis on the morning of July 20, and made my return to Rio amid brilliant sunshine, in which the glorious scenery of the bay renewed its indelible impression on my memory. In passing over the tract of low land between Raiz da Serra and the shore, partly overgrown by shrubs or small trees ten or twelve feet in height, I found them covered with masses of large flowers of the most brilliant purple hue, where ten days before not a single flower had been visible. The train halted for half a minute at a solitary half-way house, and I was able to break off a branch from the nearest plant. It belonged, as I suspected, to the family of *Melastomaceæ*, and is known to botanists as *Pleroma granulosum* of Don; but one seeing dried specimens in a European herbarium, could form no conception of the gorgeous effect of the masses of rich colour that were here displayed, outshining the splendours of the Indian rhododendrons now familiar to European eyes. I again found the same species at Tijuca; but the soil and situation were, I suppose, less favourable, and the show of bloom was neither so rich nor so abundant.

I was told that the local name of this splendid plant is *quaresma*, because it flowers in Lent, which in Brazil falls in autumn; but I afterwards ascertained that the same name is given to several other species of *Melastomaceæ* having brilliant flowers, and it seems improbable that the same species which I found bursting into flower in mid-winter should have also flowered three or four months before. The only remains of fruit that I found were dry, empty capsules that had apparently survived the preceding summer.

Although I reached Rio some time before midday, so many matters required my attention that I found it impossible to return for a fuller visit to the Botanic Garden. Mr. Corbett had kindly offered to present me to the emperor, and, if time had permitted, I should have gladly taken the opportunity of making the personal acquaintance of a sovereign who stands alone among living rulers for the extent and variety of his scientific attainments, and for the active interest he has shown in the progress of natural knowledge. Irrespective of the qualities that appeal to the sympathies of men of science, Dom Pedro is evidently one of the remarkable men of our time. His exceptional energies, physical and mental, are incessantly devoted to every branch of public affairs, and it is said that he has even succeeded in inspiring some of his subjects with a share of his own zeal. But, so far as I could learn, he cannot be said to have achieved popularity. Among indolent and listless people, indefatigable industry produces an unpleasant effect. Improvements may or may not be desirable, but they are certain to give some trouble: it would be far pleasanter to let things remain as they are. Perhaps, whenever the time comes for Brazil to be deprived of the guidance of the present emperor, its people will become sensible of the loss they have sustained.

The steamer of the Royal Mail Company was to depart on July 24, so that no time was to be lost in making my visit to Tijuca. That place lies among the hills north-west of the city, about nine hundred feet above the sea, and the distance is quite inconsiderable; but the arrangements for visitors are inconvenient. A tramway runs over the flat country to the foot of the hill, and from the terminus the remainder of the way is accomplished by carriage or omnibus. But no luggage is taken by the tramway, and this has to be forwarded on the previous day. When I reached the station, about eleven a.m. on the 21st, I had an unpleasant quarter of an hour, during which it appeared that the case containing most of my Petropolis collections was lost or mislaid. At length it was found lying in an outhouse; no omnibus was available, but I soon succeeded in hiring a carriage to convey me to Tijuca.

The country between the city and the lower slopes of the hills is covered with the villas of wealthy natives, many of them large and handsome houses, each surrounded by a garden or pleasure-ground. In these grounds the mango, bread-fruit tree, and others, with large thick leaves giving dense shade, were invariably planted; and here and there palms, of which I thought I could distinguish four or five species, gave to the whole the aspect of completely tropical vegetation. Amidst the mass of trees, it was rarely possible to get a glimpse of the exquisite scenery surrounding Rio on every side, and it was only towards the top of the hill that I gained a view of the bay. Tijuca lies on the farther, or westward, slope, nearly surrounded by forest, and consists of only a few houses, of which the chief is White's Hotel. As I afterwards learned, Mr. White, who is engaged in business in the city, was in the habit of hospitably entertaining his friends at a spot which naturally attracted frequent visits, and at length judiciously turned his house into an hotel, where a moderate number of guests find charming scenery, comparative coolness in the hot season, and far more of creature-comforts than are to be had in the hotels of Rio.

Time allowed me no more than a short stroll in the immediate neighbourhood before the hour of dinner, at which I met several intelligent and well-informed gentlemen, and amongst them three English engineers, from whom I received much information as to the country which they have made their home.

Amongst other questions discussed was that, so important to Europeans, regarding the annual visitation of yellow fever and the best method of treatment. I was especially struck by the experiences of the youngest of the party, who had come out from England a few years before to superintend some considerable new works for the drainage of Rio. For two years he lived altogether in the city, constantly requiring to go below, and sometimes remaining for hours in the main sewers. During that time he was never attacked by the fever, and no fatal cases arose among the workmen engaged in the same work. Since its completion this gentleman had been engaged on other works of a more ordinary character, and had habitually slept in the country during the hot season; but, under conditions apparently more favourable, he had been twice stricken by the fever. The first attack, which was probably slight, was at once cut short by a large dose of castor oil and aconite administered by a friend. In the following year he experienced a more serious attack, and had been treated by a doctor of good repute, mainly with tartar emetic. It appears that professors of the healing art in Brazil regulate their charges, not by the amount of time or labour which they give, but by the estimated value of the patient's life. If he survives, it is considered that the remuneration should be in the nature of salvage—a considerable percentage on the amount of his income. In the present case the young engineer had been required to pay a fee of £180. In some cases, where the doctor's demand appeared utterly unreasonable, foreigners have attempted to appeal to the tribunals, but it appears that the results of litigation have not encouraged others to resort to the protection of the law.

In answer to my inquiries, most of my informants made light of the difficulties of exploring the interior of Brazil, but they agreed in the opinion that much time must be given by any traveller wishing to break new ground. Even in the more or less fully settled provinces, the spaces to be traversed are so great, and the means of communication so imperfect, that a large margin must be left for unexpected delays. One gentleman, who had travelled far in Goyaz and Matto Grosso, assured me that he had never encountered any difficulty as to provisions. Three articles of European origin are to be found, so he assured me, at every inhabited place in the interior— Huntley and Palmer's biscuits, French sardines, and Bass's pale ale.

July 22 was a day of great enjoyment, devoted to the immediate neighbourhood of Tijuca, where objects of interest were so abundant as to furnish ample occupation for many days. I have said that the place is almost surrounded by the forest which spreads over the adjoining hills. I now learned that less than fifty years before, at a time when coffee-planting in Brazil became a mania, and was counted on as everywhere a certain source of wealth, the aboriginal forest which covered the country was completely cleared, and coffee-planting commenced on the largest scale. Experience soon proved that the conditions either of soil or climate were unfavourable, and after a few years the land was again abandoned to the native vegetation. About thirty-five years had sufficed to produce a new forest, which in other lands might be supposed to be the growth of centuries. The trees averaged from two to three feet in diameter, and many were at least seventy feet in height. One of the largest is locally called *ipa*; it belongs to the leguminous family, has a trunk nearly quite bare, and the upper branches bore masses of cream-coloured flowers; but, finding it impossible to obtain flower or fruit, I have been unable to identify it. The vegetation here appeared to be even more luxuriant than that of Petropolis, and to indicate a somewhat higher mean temperature. The proportion of tree ferns was decidedly greater, and a good many conspicuous plants not seen there were here abundant. Some of these, such as *Bignonia venusta*, *Allamanda*, etc., may have strayed from the gardens; but many more appeared to be certainly indigenous. Of flowering plants the family of *Melastomaceæ* was decidedly predominant, and within a small area I collected fifteen species, eight of which belonged to the beautiful genus *Pleroma*. One of these (*P. arboreum* of Gardner) is a tree growing to a height of forty feet; but the species of this family are more commonly shrubs not exceeding ten or twelve feet in height.

I was unfortunately not acquainted at that time with the observations made near Tijuca by Professor Alexander Agassiz, which appear to him to give evidence of glacial action in this part of Brazil. It would be rash, especially for one who has not been able to examine the deposits referred to, to controvert conclusions resting on such high authority; but I may remark that the evidence is confessedly very imperfect, and that the characteristic striations, either on the live rock or on the transported blocks, which are commonly seen in the theatre of glacial action, have not been observed. I lean to the opinion that the deposits seen near Tijuca are of the same character as those described by M. Liais* as frequent in Brazil. The crystalline rocks are of very unequal hardness, and while some portions are rapidly disintegrated, the harder parts resist. The disintegrated matter is washed away, and the result is to leave a pile of blocks of unequal dimensions lying in a confused mass.

On the following day, my last in Brazil, one of my new acquaintances was kind enough to guide me on a short excursion in the forest, which enabled me to

---

* See his valuable work, "Climats, Géologie, Faune et Géographie Botanique du Brésil."

approach one of the giants of the vegetable kingdom. At the time of the clearing of the aboriginal forest two great trees were spared. One of these had been blown down some years before my visit, and but one now remained. It was easily recognized from a distance, as it presented a great dome of verdure that rose high above the other trees of the forest. The greater part of the way was perfectly easy. A broad track, smooth enough to be passable in a carriage, has been cleared for a distance of many miles over the forest-covered hills.

Following this amid delightful scenery, we reached a point scarcely two hundred yards distant from the great tree. I had already learned that even two hundred yards in a Brazilian forest are not very easily accomplished, but I was assured that a path had been cut a year or two before which allowed easy access to the foot of the tree. We found the path, but it was soon apparent that it had been neglected during the past season, and in this country a few months suffice to produce a tangle of vegetation not easily traversed.

When at length we effected our object, we found ourselves at the base of a cylindrical column or tower, with very smooth and uniform surface, tapering very slightly up to the lowest branch, which was about eighty feet over our heads. We measured the girth, and found it just twenty-nine feet at five feet from the ground. It is needless to say that I could form no conjecture as to the species, or even the family, to which this giant belongs, as I was quite unable to make out the character of the foliage. While near to it we could form no guess as to the height; but my companion, whose profession made him used to accurate estimates, and who had observed it from many points of view, reckoned the height at between 180 and 200 feet. I had not then seen the giant conifers of western North America, but, excluding the two Sequoias, I have not found any single tree to equal this. In the valleys of the Alleghany Mountains in Tennessee, I have indeed beheld not unworthy rivals. The *Liriodendron* there sends up a stem more than seven feet in diameter, and frequently exceeds 150 feet in height.

To diminish my regret at quitting this beautiful region, the morning of July 24 broke amid dark clouds and heavy rain, which continued till the afternoon. I had counted on enjoying a few hours in Rio before my departure, but, that being impossible, I went directly from Tijuca to the landing-place, and thence on board the steamer of the Royal Mail Company, which was to take me back to England. This was the *Tagus*, and I had much pleasure in finding her under the command of Captain Gillies, with whom I had made the voyage from Southampton to Colon. In the afternoon we slowly steamed out of the bay. Its glories were veiled, heavy clouds rested on the Organ Mountains; but the Corcovado and the other nearer summits appeared from time to time, and the last impression was that of fleeting images of beauty the like of which I cannot hope again to behold.

The course for steamers from Rio Janeiro to England is as nearly as possible direct. The coast of Brazil from Rio to Pernambuco runs from south-

south-west to north-north-east, in the same direction that leads to Europe; and from the headland of Cabo Frio to the entrance of the English Channel at Ushant, a distance of about 72° of latitude and 38° of longitude, the helm is scarcely varied from the same course. It is somewhat remarkable that in so long a voyage, in which one passes from the Tropic of Capricorn to the region of the variable anti-trade winds of the northern hemisphere, it not very rarely happens, as I was assured by our experienced captain, that north-north-east winds are encountered throughout the entire distance. This was nearly verified in the present case. For comparatively short periods the wind shifted occasionally to the north and north-west; more rarely, and at brief intervals, light breezes from the south and south-east were experienced; but the north-east and north-north-east winds predominated, even on the Brazilian coast, until we reached the latitude of Lisbon.

It is an admitted fact in meteorology, that the trade winds of the northern are—at least in the Atlantic—stronger than those of the southern hemisphere; but at the winter season of the south, the south-east trade winds prevail in the equatorial zone, and are not rarely felt as far as eight or even ten degrees north of the equator. But in investigating the extremely complex causes that determine the direction of air currents, and especially those slight movements that make what is called a breeze, it is difficult to trace the separate effect of each agent. The neighbourhood of a coast constantly brings local causes into play, and it may well be that the rapid condensation of large masses of vapour, such as occurs at each heavy fall of rain, may determine temporary currents in the air in directions opposed to the general and ordinary march of the winds. Irrespective, however, of any such local causes, we must bear in mind the general tendency of air currents towards motion of a circular or spiral character. When we meet a breeze blowing in a direction contrary to that which ordinary experience leads us to expect, we must not forget the possibility that it may be a portion of the ordinary current which has formed an eddy. The main facts of meteorology are now well established, but the local deviations may give room for prolonged study.

Although I knew that the delay at both places would be short, I looked forward with much interest to the prospect of landing at Bahia and Pernambuco. The latter place especially is known to be the chief mart for the natural productions of Equatorial America. Skins of animals, birds living and dead, gorgeous butterflies and shells, are easily procurable; and a mere visit to the fish and vegetable markets is sure to make a visitor acquainted with objects of interest. But my expectations were doomed to disappointment.

We reached Bahia on the morning of July 27. The city stands on a point of land north of the entrance to an extensive bay, called by the Portuguese Bahia de Todos Santos, and the proper name of the city is São Salvador; but the concurrent practice of foreigners has established the name now in general use. The steamer lay in the roadstead nearly a mile from the shore, and the heavy boats, carrying some passengers for Europe, moved slowly as they pitched to and fro in the swell of the sea. Just as they came alongside, rain suddenly burst in a torrent from the clouds, which had formed in the course of a few minutes. For the first time in my journey, I regretted the omission to have supplied myself with a waterproof cloak. A thorough wetting in tropical countries usually entails an attack of fever, and for that I was not prepared; so, along with two or three other passengers who wished to go ashore, I remained in the main deck. The rain ceased, and there was an interval of sunshine; but the torrential showers were renewed two or three times before we resumed our voyage in the afternoon.

I have already noticed the contrast that exists between the winter and summer climate of this part of Brazil and that of Rio and the southern provinces. In the latter the rainy season is in summer, while nearer the equator, although no season can be called dry, the chief rainfall occurs in winter—that is to say, in the season when the sun is farthest from the zenith. While passing through the equatorial zone, when intervals of bright weather alternated with extremely heavy rain, I frequently consulted the barometer, but was unable to trace the slightest connection between atmospheric pressure and rainfall, the slight oscillations observed being due to the diurnal variation everywhere sensible in the tropics.

The temperature on this part of the coast was only moderately warm, varying from 76° to 78° Fahr. on this and the following day, when we called at Maceio, a place of increasing commercial importance. Our stay was so short that no one attempted to go ashore, although the weather was favourable. Several whales were seen both on the 27th and 28th, but I failed to ascertain to what species they belonged.

On the evening of the 28th we experienced a decided rise of temperature; three hours after sunset the thermometer still stood at 81° Fahr., and, with two remarkable intervals, it did not fall below 80° during the following eight days. During that time my attention was often directed to the physiological effects of heat on the human economy, and both my own experience and the conflicting testimony of travellers lead me to conclude that there are many facts not yet satisfactorily explained.

On the enfeebling effect of moist tropical climates there is a general agreement, both as to the fact and the chief cause; but, as I have remarked in a preceding page, the circumstance that this is little or not at all experienced at sea is apparently anomalous. With regard to the direct effect of the sun's rays on the surface of the body, and especially in the production of sun-stroke, the evidence of scientific travellers is conflicting, and the explanations offered are by no means satisfactory. On the one hand, it is asserted on good authority that in the equatorial zone the direct effect of the sun is far greater than it is in Europe at the same elevation above the horizon. The rapid reddening and blistering of the skin where exposed, and sun-stroke from exposure of the head, are said to be the ordinary effects. Being extremely sensitive to solar heat, I have always carefully protected my head, and have avoided rash experiments. Of the reddening and blistering of the skin I have had very frequent experience in Europe, upon the Alps and other mountains; but I observed none but very slight effects of this kind in the tropics, even with a nearly vertical sun, either on land or while at sea. Dr. Hann[*] cites many statements on the subject. In the West Indies cases of sun-stroke are rare, and the inhabitants expose themselves without danger. In nearly all parts of British India, as is too well known, the danger of exposing the head to the sun is notorious, and the same is certainly true of most parts of tropical Africa.

The most obvious suggestion is that, inasmuch as dry air absorbs less of the solar heat than air charged with aqueous vapour, the injurious effects should be more felt in dry climates than in damp ones. But, so far as what is called sun-stroke is concerned, the balance of evidence is opposed to this conclusion. Sir Joseph Fayrer, who has had wide experience in India, expressly asserts that the hot dry winds in Upper India induce less cases of sun-stroke than the moist though cooler climate of Bengal and Southern India. Dr. Hann quotes Borius for a statement that in Senegambia the rainy season is that in which sun-stroke commonly occurs, while he further asserts that on the Loango coast, in very similar climatal conditions, the affection is almost unknown, and that Europeans even expose the head to the sun with impunity.

My own conclusion, fortified by that of eminent

[*] "Klimatologie," p. 382.

authorities, is that the phenomena here discussed are of a very complex nature; that different physical agencies are concerned in the various effects produced on the body; and that most probably there are many different pathological affections which have been classed together, but which, when more fully studied, will be recognized as distinct.

In the first place, I apprehend that the action of the sun which causes discolouration and blistering of the skin has no relation to that which causes sun-stroke. It is a local effect confined to the surfaces actually exposed, and, if it could be accurately registered, would serve the purpose of an actinometer, depending as it does on the amount of radiant heat reaching the surface in a unit of time.

Sun-stroke proper is, I believe, an affection of the cerebro-spinal system arising from the overheating of those parts of the body. It is by no means confined to the tropics, or to very hot countries, as many cases occur annually in Europe, and still more frequently in the eastern states of North America.

Nearly allied to sun-stroke, but perhaps sufficiently different to deserve separate classification, are those attacks which some writers style cases of thermic fever, which arise mainly in places where the body is for a continuance exposed to temperatures exceeding the normal amount of the human body. In producing thermic fever, it would appear that the depressing effect of a hot moist climate acts powerfully as a predisposing cause, and such cases not uncommonly arise where there has been no exposure whatever to the direct rays of the sun.

It is easy to understand that, as a general rule, seamen are less exposed than other classes to any of the injurious effects of heat, but it is remarkable that they should enjoy complete exemption. Cases are not very uncommon among seamen going ashore in hot countries, but I have not found a well-authenticated case of sun-stroke arising on board ship; and cases of thermic fever in the Red Sea usually arise in the engine-room of a steamer rather than among the men on deck.

On the morning of July 29 we reached Pernambuco, to which I had looked forward as the last Brazilian city that I was likely to see. It had been described to me as the Venice of South America, and the comparison is to a slight extent justified by its position on a lagoon of smooth water, separated from the open roadstead by a coral reef several miles in length. It enjoys the further distinction, unusual in a place within eight degrees of the equator, of being remarkably healthy. But on this occasion fortune was against me.

No doubt for some sufficient reason, we did not enter the rather intricate passage leading inside the reef, but lay to in rough water outside. For a short time the scene was brilliant. The hot sun beat down on the deep blue water, and lit up the foam on the crests of the dancing waves, and the sky overhead showed such a pure azure that one could not suppose the air to be saturated with vapour. Before long boats were seen approaching, tossed to and fro in the broken water; but before they drew near, heavy clouds had gathered in the course of a few minutes, and a torrent of water was discharged such as I had never experienced except in passing under a waterfall. As each boat came alongside, a seat was let down from the upper deck, and the passengers were hoisted up in turn, those who had not efficient waterproofs being as thoroughly drenched as if they had been dipped in the sea. Four or five times during the day the sky cleared, the blazing sun returned, and the decks were nearly dry, when another downpour of torrential rain drove us all to seek shelter, each shower lasting only from ten to fifteen minutes.

During the hotter hours of the day a rather strong breeze set in towards the shore, and I have no doubt that it is to its full exposure to this ordinary sea-breeze that the city owes its comparative healthiness. It was interesting to watch the manœuvres of the *catamarans*, in which the native fishermen were pursuing their avocations. This most primitive of sea-craft is formed of two or three logs well spliced together, with some weight to serve as ballast fastened underneath. In the forepart a stout stick some ten feet long stands up as a mast and supports a small sail, and amid-ships a short rail, supported on two uprights, enabled the two men who form the crew to hold on when much knocked about by the waves. A single paddle seems to serve as a rudder, but it is not easy to understand how such a rude substitute for a boat is able to work out to sea against the breeze which commonly sets towards the shore.

A large proportion of the steerage passengers who came on board at Bahia and Pernambuco were Portuguese returning to their native country after a residence, either as artisans or as agricultural settlers, in Brazil. My command of the language is unfortunately so limited that I failed to extract from these fellow-passengers any interesting information. With scarcely an exception, each carried at least one parrot, usually intended for sale at Lisbon, where it appears that they are in some request. Comparatively high prices are given for birds that freely simulate human speech.

We were under steam in the afternoon of the 29th, and soon lost to view the South American continent. On the following day the barometer for the first time showed the diminution of pressure which is normally found in the equatorial zone. Between nine a.m. and four p.m. the ship's mercurial barometer fell about a quarter of an inch from 30·30 to 30·06 inches, and my aneroid showed nearly the same amount of difference. It must be remembered, however, that nearly one-half of the effect (at least one-tenth of an inch) must be set down to the daily oscillation of the height of the barometer, which so constantly occurs within the tropics, the highest pressures recurring at ten a.m. and ten p.m., and the lowest about four p.m. and four a.m.

I carried with me on this journey only a single aneroid barometer, an excellent instrument by Casella, whose performance was very satisfactory, and which in a very short time returned to its normal indication after exposure to diminished pressure in the Andes; but it had the defect, which, so far as I know, is common to the aneroid instruments by the best makers, that the temperature at which the scale is originally laid down by comparison with a standard mercurial barometer is not indicated on the face of the instrument. Assuming that the aneroid is compensated for variations of temperature, and I have found this to be the case within ordinary limits in good instruments, there remains the question to what height of mercury at what temperature a given reading of the aneroid corresponds. For scientific purposes it is customary to reduce the reading of the mercurial barometer to the temperature of the freezing-point of water, and it is often supposed that the aneroid reading corresponds to that figure. But we may feel pretty confident that the maker, in laying down the scale, did not work in a room at freezing-point. I have been accustomed to assume 15° Cent., or 59° Fahr., as about the probable temperature with instruments made in our climate.

In the present case, the barometer-reading of 30·06 inches at the temperature of 84° Fahr. would (neglecting the small correction for capillarity) be reduced by about fourteen-hundredths of an inch, in order to give the correct figure at freezing-point; but for comparison with an aneroid, supposed to have been laid down at 59° Fahr., the correction would be a fraction over seven-hundredths of an inch. As a matter of fact, my aneroid marked at four p.m. 29·89 inches, or, allowing for the correction, just one-tenth of an inch less than the ship's mercurial barometer, and, as I believe, was more nearly correct.

As the sun was declining on the evening of July 30, we sighted the remarkable island of Fernando Noronha. It lies about four degrees south of the equator, and more than two hundred miles from the

FERNANDO NORONHA.

nearest point of the Brazilian coast. The outline is singular, for the rough hills which cover most of the surface terminate at the western end of the island in a peak surmounted by a column, in the form of a gigantic lighthouse, which must rise over a thousand feet above the sea-level.* Although Darwin passed some hours on the island in 1832, it remains to the present day one of the least known of the Atlantic islands, so far as regards its natural productions. A fellow-passenger who had landed there assured me that he had found granite; but I have no doubt that the island is exclusively of volcanic origin, for such is the opinion of the few scientific men who have visited it.

The island has been converted by the Brazilian Government into a convict station, and in consequence access by strangers has become very difficult. Such information as we possess is mainly to be found in Professor Moseley's account of the voyage of the *Challenger*. On landing there with Sir G. Nares, he at first obtained permission from the governor to visit the island and to collect natural objects; but the permission was very soon retracted, and he was unable to obtain specimens of several singular shrubs that abound and give the island the appearance of being covered with forest.

Now that the attention of naturalists has been directed to the especial interest attaching to the fauna and flora of oceanic islands, and their liability to

---

* Darwin's estimate of the height was one thousand feet, while Professor Moseley gives double that amount. I incline to think that the lower figure is nearer to the truth.

extinction owing to competition from species introduced by settlers, it may be hoped that the exploration of this small but remarkable island will before long be undertaken by a competent naturalist. For that purpose it would be, in the first place, necessary to obtain the permission of the Brazilian Government, and to secure the means of existence during a stay of ten or twelve days on the island. The most effectual means would be through direct personal application to the emperor, who is well known to take a lively interest in all branches of natural science.

With the thermometer standing about 82°, the passengers naturally preferred the upper deck to the close air of the saloon, and were resting in their ship-chairs between nine and ten p.m., when suddenly there came an outburst of coughing and sneezing, followed by demands for muffling of every kind. There was no sensible movement in the air, but I found that the thermometer had fallen to 79° Fahr., and there was a feeling of chilliness which was not easily explained by that slight fall of temperature.

The mystery was explained on consulting the chief officer, who throughout the voyage paid much attention to the temperature of the sea. Since leaving Pernambuco, the thermometer in buckets brought up from the surface had varied only between 82° and 83°. On this evening we had abruptly encountered a relatively cold current, with a temperature somewhat below 76°, and the effect of being surrounded by a body of cool water when the skin was in the condition usual in the tropics was felt by nearly all the passengers.

With slight variation, this comparatively cool current must have extended over a large area on both sides of the equator, as the temperature of the water remained nearly the same for about forty-eight hours.

Throughout the voyage from Brazil to Europe, I was fortunate in enjoying the society of a man of remarkable intelligence, who has been a diligent and accurate observer of nature in a region still imperfectly known. M. Georges Claraz, by birth a Swiss, belonging to a family of small proprietors in the Canton of Fribourg, had gone out as a young man to improve his fortune in South America. He had received a fair scientific education, having followed the lectures of the eminent men who have adorned the Polytechnic School at Zurich; but, what is much more rare, he appeared to have retained everything that he had ever learned, and to have had a clear perception of the scientific value of the observations that a stranger may make in a little-known region. After passing some time in the state of Entrerios, he had settled at Bahia Blanca, close to the northern border of Patagonia. He had established friendly relations with the Indians, and made frequent excursions in the interior of Patagonia and southward as far as, and even beyond, the river Chubat.

During the entire time, although engaged in the work of a settler, M. Claraz seems to have made careful notes of his observations—on the native Indians and their customs; on the indigenous and the domestic animals; on the plants and their uses; on the mineral structure of the country, not omitting to take specimens of the mud brought down by the different rivers; and on general physics. Of his large collections I trust that the greater part have safely reached Switzerland. A considerable collection of dried plants, sent home while he resided at Bahia Blanca, was unfortunately lost. He was good enough, after his return, to send me a smaller collection remaining in his hands, of which I gave an account in the *Journal of the Linnæan Society* for 1884.

As I trust that the great store of information collected by M. Claraz will before long be given to the world, I should not wish to anticipate the appearance of his work, but I may say that among many interesting particulars, several of which I noted at the time, I was especially struck by the evidence collected among the Indians, which seemed to prove that the *Glyptodon* survived in Patagonia down to a comparatively recent period, and that the tradition of its presence is preserved in the stories and songs of the natives.

Early on July 31 we passed the equator, but it was not till ten p.m. on the following day that we escaped from the area of cool water and found the ordinary equatorial temperature of 82·5°. During the three following days the weather was hot and relaxing, the thermometer ranging by day between 84° and 85°. For some hours on the 2nd of August the wind came from south-south-east, but before evening it backed to west, and blew from that point rather freshly at night. On the following day we appeared to have met the north-east trade wind, which was, however, a gentle breeze, and occasionally veered to the north-west.

In the afternoon of August 4 we made out the picturesque outline of the Cape Verde Islands, and before sunset entered the channel between St. Vincent and St. Antão, finally dropping anchor for the night in the outer part of the fine harbour of St. Vincent. Having been selected as a coaling station, this has become the chief resort of steamers plying between Europe and the Southern Atlantic, and we were led to expect that the operation would take up great part of the following day. Here a fresh disappointment awaited me. I had confidently reckoned upon spending several hours ashore, and seeing something of the curious vegetation of the island, which includes a scanty representation of tropical African types, with several forms allied to the characteristic plants of the Canary Islands.

I had not duly taken account of the perverse temper of the officers of health, whose chief object in life seems everywhere to be to make their authority felt by the needless annoyance they cause to unoffending fellow-creatures. We had left Rio with a clean bill of health; not a single case of yellow fever had occurred for months before our departure; but Brazil is regarded as permanently "suspected," and quarantine regulations were strictly enforced in our case.

I am far from believing that in certain conditions, and as regards certain diseases, judicious quarantine regulations may not be effective; but, reckoning up all the loss and inconvenience, and the positive damage to health, arising from the sanitary regulations now enforced, I question whether it would not be better for the world if the system were entirely abolished.

The view of St. Vincent, backed by a bold and stern mountain mass, on which scarcely a trace of vegetation is visible from a distance, was for some time sufficiently interesting; but as the day wore on, and the sun beat down more fiercely, life on board became less agreeable. To keep out the penetrating coal dust all the ports were closed, and, with the thermometer at 90°, the air below was stifling, and the passengers generally preferred to remain on deck, and breathe the hot air mixed with the coal dust that arose from the open bunkers.

I offered two of the boatmen who hung about the ship three milreis if they would land on an uninhabited part of the bay, which I pointed out to them, and collect for me every plant they found growing, and I was well pleased when, after two or three hours, they returned with a respectable bundle of green foliage. Under the vigilant eyes of the officers of health the specimens were hauled up to the deck, while the three dollars were thrown into the boat. It is remarkable that coin is nowhere supposed to convey contagion.

When I came to examine it, I found to my disgust that the bouquet included only the leaves of two species, with no trace of flower or fruit. One was most probably *Nicotiana glauca*, introduced from tropical America; the other a leguminous shrub, possibly a *Cassia*, but quite uncertain.

The rest of the passengers spent most of the day in bargaining with the hucksters who flocked round the ship. Ornaments made from palm leaves, sweetmeats of very suspicious appearance, photographs, and tobacco in various forms, were the chief articles of traffic, and the main object seemed to be to prolong the chaffering and bargaining over each article so as to kill as much time as possible. More attractive in appearance were the tropical fruits, of which those suitable to a dry climate grow here in perfection. In spite of persevering efforts, I have never developed much appreciation of the banana as an article of diet, but I thought those obtained here much the best that I have anywhere eaten.

General satisfaction was felt when, the work of coaling being finished, the ship was again in motion, with her head set towards Europe. On returning to the channel between the islands, and still more when we had got well out to sea, we encountered a rather strong breeze right ahead, which with varying force continued for the next four days. This was, of course, the regular trade wind of the North Atlantic, and had the agreeable effect of lowering the temperature, which at once fell to 78°. Along with the trade wind, the sea-current apparently travels in the same direction. It is certain that the temperature of the water was here much lower. Before reaching St. Vincent we found it between 80° and 81° Fahr., while after leaving the islands it had fallen to 74°. This temperature remained nearly constant for three days, but on the evening of the 9th, in about 27° north latitude, we abruptly encountered another current of still cooler water, in which the thermometer fell to 69°.

The force of the wind never, I think, exceeded

what seamen describe as a fresh breeze, but it sufficed to cause at times considerable disturbance of the surface; and on the afternoon of the 6th we shipped some heavy seas, so that it was found expedient to slacken speed for a time.

I have alluded in a former page to the ordinary observation that in the track of the trade winds the breeze usually falls off about sunset. It is more difficult to account for the opposite phenomenon, which we experienced on three successive evenings from the 7th to the 9th of August, when the force of the wind increased in a marked degree after nightfall.

I was also struck by the fact that the temperature of the air throughout the voyage from St. Vincent to the mouth of the Tagus seemed to be unaffected either by the varying force of the wind or by the fall in surface-temperature of the sea, to which I have above referred. On board ship in clear weather it is very difficult to ascertain the true shade-temperature when the sun is much above the horizon, but the observations made at sunrise and after nightfall from the evening of the 5th to the morning of the 11th varied very slightly, the utmost range being from $77\cdot5°$ to $73°$.

Some points in the Canary Islands are often visible in the voyage from Brazil to Europe, especially the lofty peak of Palma; but we passed this part of the course at night, and nothing was seen. As we drew near to Europe, the wind, through keeping the sam direction, gradually fell off to a gentle breeze, and the surface of the water became glassy smooth, heaving gently in long undulations. The relative effect of smooth or rough water on the speed of steamers is remarkable, and was shown by the fact that during the twenty-four hours ending at noon on the 11th of August the *Tagus* accomplished a run of 295 knots, while three days before, with only a gentle breeze but rougher water, the run to noon was only 240 knots.

Early in the afternoon of the 11th, the Rock of Lisbon at the mouth of the Tagus was distinctly visible, and we slowly entered the river and cast anchor at the quarantine station below Belem. Our captain, after the experience of St. Vincent, did not expect to obtain pratique at Lisbon, and with more or less grumbling the passengers had made up their minds to remain on board, when, after a long deliberation, the unexpected news, "admitted to pratique," was rapidly spread through the ship, and we moved up to the anchorage opposite the picturesque old tower of Belem, which the true mariner must always regard as one of his holy places. It marks the spot wherefrom Vasco de Gama and his companions, after a night spent in prayer in the adjoining chapel, embarked on their memorable voyage, and here, after years of anxious uncertainty, King Manuel greeted the survivors on their return to their country.

The sun was sinking when such passengers as wished to see something of Lisbon took the opportunity for going ashore, while others, like myself, preferred to remain on board. Hoping to receive letters at the post-office, I landed early next morning, and found a tramcar to carry me to the centre of the town. Early hours are not in much honour at Lisbon. I found the post-office closed, and, after several vain efforts, was informed that letters could not be delivered until ten o'clock, the precise hour fixed for our departure from the anchorage at Belem.

The voyage from Lisbon along the coasts of Portugal and Galicia is usually enjoyed, even by fair-weather sailors. The case is often otherwise with the Bay of Biscay, but on this occasion there was nothing of which the most fastidious could complain. I have sometimes doubted whether injustice has not been done to that much-abused bay, which, in truth, is not rightly so called by those bound from the north to the coast of Portugal. It is simply a part of the Atlantic Ocean, adjoining the coast of Europe between latitudes $43° 46'$ and $48° 28'$. I have not been able to ascertain that the wind blows harder, or that the sea runs higher there than elsewhere in the same latitudes, and am inclined to rank the prejudice against that particular tract of sea-water among vulgar errors.

The adventurer who has attempted to open up a trade with some distant region is accustomed, as he returns home, to count up the profits of his expedition; and in somewhat the same spirit the man who pursues natural knowledge can scarcely fail to take stock of the results of a journey. It is his happy privilege to reckon up none but gains, and those of a kind that bring abiding satisfaction. He may feel some regret that outer circumstance or his own shortcoming have allowed opportunities to escape, and lessened the store that he has been able to accumulate; but as for the positive drawbacks, which seemed but trivial at the time, they absolutely disappear in the recollection of his experiences. Thinking of these things as the journey drew to a close, I could not help feeling how great are the rewards that a traveller reaps, even irrespective of anything he may learn, or of the suggestions to thought that a voyage of this kind cannot fail to bear with it. How much is life made fuller and richer by the stock of images laid up in the marvellous storehouse of the brain, to be summoned, one knows not when or how, by some hidden train of association—shifting scenes that serve to beautify many a common and prosaic moment of life!

Often during this return voyage my thoughts recurred to an article in some periodical lent to me by my kind friends at Petropolis, wherein the writer, with seeming gravity, discussed the question *whether life is worth living*. My first impression, as I well remember, was somewhat contemptuous pity for the man whose mind could be so profoundly diseased as even to ask such a question, as for a soldier who, with the trumpet-call sounding in his ear, should stop to inquire whether the battle was worth fighting. When one remembers how full life is of appeals to the active faculties of man, and how the exertion of each of these brings its correlative satisfaction; how the world, in the first place, needs the daily labour of the majority of our race; how much there is yet to be learned, and how much to be taught to the ignorant; what constant demand there is for the spirit of sympathy to alleviate suffering in our fellows; how much beauty exists to be enjoyed, and, it may be, to be brought home to others;—one is tempted to ask if the man who halts to discuss whether life is worth living can have a mind to care for truth, or a heart to feel for others, or a soul accessible to the sense of beauty.

Recurring to the subject, as I sometimes did during the homeward voyage, it seemed to me that I had perhaps treated the matter too seriously, and that the article I had read was an elaborate hoax, by which the writer, while in truth laughing at his readers, sought merely to astonish and to gain repute as an original thinker. However the fact may be, when taken in connection with the shallow pessimism which, through various channels, has of late filtered into much modern literature, there does appear to be some real danger that the disease may spread among the weaker portion of the young generation. A new fashion, however absurd or mischievous, is sure to have attractions for the feebler forms of human vanity. It is true that there is little danger that the genuine doctrine will spread widely, but the mere masquerade of pessimism may do unimagined mischief. The better instincts of man's nature are not so firmly rooted that we should wish to see the spread of any influence that directly allies itself with his selfish and cowardly tendencies.

To any young man who has been touched by the contagion of such doctrines, I should recommend a journey long enough and distant enough to bring him into contact with new and varied aspects of nature and of human society. Removed from the daily round of monotonous occupation, or, far worse, of monotonous idleness, life is thus presented in larger and truer proportions, and in a nature not quite worthless some chord must be touched that will stir the springs of healthy action. If there be in truth such beings as genuine and incurable pessimists, the stern believer in progress will be tempted to say that the sooner they carry out their doctrine to its logical result the better it will be for the race. Their continued existence, where it is not merely useless, must be altogether a mischief to their fellow-creatures.

On the morning of the 16th of August, all but completing five months since I quitted her shores, the coast of England was dimly descried amid gusts of cold wind and showers of drizzling rain. My winter experiences in the Straits of Magellan were forcibly recalled to my mind, and I felt some partial satisfaction in the seeming confirmation of the conclusion which I had already reached—that the physical differences between the conditions of life in the northern and southern hemispheres are not nearly so great as has generally been supposed.

APPENDIX A.

ON THE FALL OF TEMPERATURE IN ASCENDING TO HEIGHTS ABOVE THE SEA-LEVEL.

THE remarkable features of the climate of Western Peru referred to in the text seem to me to admit of a partial explanation from the local conditions affecting that region. The most important of these are the prevalence of a relatively cold oceanic current, and of accompanying southerly breezes along the Peruvian coast. These not only directly affect the temperature of the air and the soil in the coast-zone, but, by causing fogs throughout a considerable part of the year, intercept a large share of solar radiation. It has been found in Northern Chili, some fifteen degrees farther south than Lima, but under similar climatal conditions, that, although the land rises rather rapidly in receding from the coast, the mean temperature increases with increasing height for a considerable distance. It is stated on good authority[*] that at Potrero Grande, a place about fifty miles distant, and 850 metres above the sea, the mean annual temperature is higher by $2\cdot 5°$ C. than at Copiapò, or at the adjoining port of Caldera. It is probable that in the valley of the Rimac the mean temperature at a height of 1000 metres is at least as high as it is at Lima. Taking the mean temperature of the lower station at $19\cdot 2°$ C., and that of Chicla at $12\cdot 2°$ C., that would give a fall of $7°$ for a difference of level of 2724 metres, or an average fall of $1°$ for 387 metres, instead of $1°$ for 512 metres, as given in the text.

A further peculiarity in the climate, which tends to diminish

---

[*] I borrow this statement from the excellent "Lehrbuch der Klimatologie," by Dr. Julius Hann. Stuttgart, 1883.

below the normal amount the rate of decrease of temperature, is the comparative absence of strong winds, and the feebleness of the sea-breezes which are usually so conspicuous in the tropics. For reasons that will be further noticed, the fall in temperature in ascending mountain ranges is largely due to currents of air carried up from the lower region. In mountain countries an air-current, encountering a range transverse to its own direction, is mechanically forced to rise along the slopes, and thus raises large masses of air to a higher level; the same effect in a less degree occurs with isolated peaks. But in the Peruvian Andes, as well as in many other parts of the great range, although storms arise from local causes on the plateau, westerly winds from the ocean are infrequent and feeble; and the sea-breezes, due to the heating of the soil by day, much less sensible than usual in warm countries.

Making full allowance for the operation of the two causes here specified, it yet appears that the difference of temperature between the coast and the higher slopes of the Peruvian Andes is exceptionally small. It is not merely due to the abnormal cooling of the coast-zone, but to the exceptionally high temperature found in the zone ranging from 3500 to 4000 metres. I should not have attached much importance to the few observations of the thermometer that I was able to make during a hurried visit, if the conclusion which they suggest had not been strongly confirmed by the character and aspect of the vegetation.

When I found that the table given by Humboldt, which has been copied and adopted by so many writers on physics, in which the mean temperature at a height of 2000 toises, or 3898 metres, in the Andes of Ecuador, close to the equator, is set down at $7°$, while at Chicla, thirteen degrees of latitude south, at a height less only by 174 metres, there is reason to believe that we find a mean annual temperature of not less than $12°$, I was led to enter more fully into the subject.

The result of somewhat careful study has been to convince me that, while the physical principles involved in the attempt to discover the vertical distribution of temperature in the atmosphere prove the problem to be one of extreme complexity, the results hitherto obtained from observation are altogether insufficient to guide us to an approximate law of distribution. I may remark that the problem has not merely a general interest in connection with the physics of the globe, but has a direct bearing on two practical applications of science. The observations of the astronomer and the surveyor require a knowledge of the amount of atmospheric refraction, by which the apparent positions of the heavenly bodies, or of distant terrestrial objects, are made to differ from the true direction ; and to determine accurately the amount of refraction we should know the temperature of the successive strata of air intervening between the observer and the object. In determining heights by means of the barometer, or any other instrument for measuring the pressure of the air, it is equally necessary for accuracy to know the variations of temperature in the space between the higher and the lower station.

Three different opinions have prevailed among physicists as to the law, or supposed law, of the rate of variation of temperature in ascending from the sea-level. The simplest supposition, and the most convenient in practice, is that the fall of temperature is directly proportional to the height, and this has been adopted in several physical treatises. In English works the rate has been stated at a fall of $1°$ Fahr. for 300 feet of ascent, and by French writers the not quite equivalent rate of $1°$ C. for 170 metres has been adopted. The formula proposed by Laplace for the determination of heights from barometric observations, which has been very generally adopted by travellers and men of science, implicitly assumes that the rate of decrease of temperature is more rapid as we ascend to the higher regions than it is near the sea-level, and this opinion was explicitly affirmed by Biot in his memoirs on atmospheric refraction. A third hypothesis may be said to have originated when, in 1862, Mr. Glaisher made his report of the results of the famous balloon ascents effected by him and Mr. Coxwell,[*] and among others exhibited a table showing the average decline of temperature corresponding to each successive thousand feet increase of elevation from the sea-level to a height of 29,000 English feet.

As Mr. Glaisher's tables showed a gradual decline in the rate of fall of temperature with increasing height, they clearly did not accord with the ordinary assumption of an uniform rate,

---

[*] See *Reports of the British Association for the Advancement of Science* for 1882, pp. 451-453.

and still less with the hypothesis of Laplace and Biot. In February, 1864, Count Paul de St. Robert, of Turin, communicated to the *Philosophical Magazine* a short paper, in which he showed the incompatibility of Mr. Glaisher's results with the ordinary formulæ for the reduction of barometric observations, and proposed a new formula based on a law of decrement of heat based upon Mr. Glaisher's tables. In the following June, M. de St. Robert published in the same journal a further paper, in which, still accepting Mr. Glaisher's results as accurate, he investigated the subject in a masterly manner, as well with reference to the measurement of heights, as in its connection with the determination of the amount of atmospheric refraction. The formula proposed by M. de St. Robert, and the tables subsequently published by him for its adaptation to use, appearing to be at once the most accurate and the most convenient, have been adopted by myself and by many other travellers;[*] but it is evident that their value depends on the correctness of the results, above referred to, deduced by Mr. Glaisher, and their conformity with observation in mountain countries.

Before we inquire into the conclusions to be drawn from observation, it may be well to point out how incomplete is our knowledge of the physical agencies which regulate the distribution of temperature in the atmosphere.

The primary source of temperature is solar radiation, and its effect at any given point on the earth's surface depends on the absolute amount of heating power in the sun's rays, irrespective of absorption, commonly designated the *solar constant*, and on the proportion of heat which is lost by absorption in passing through the atmosphere. The temperature of the air at any point will, in the first place, depend on the amount of solar radiation and of heat radiated from terrestrial objects directly absorbed, and next on the heating of the strata near the surface by convection. The amount of heat received from the sun, directly or indirectly, varies of course with the sun's declination

---

[*] It is remarkable that there is no reference to the investigations of M. de St. Robert, and the formula deduced from them, in the article on the "Barometrical Measurement of Heights," in the new edition of the *Encyclopædia Britannica*.

at the time, and the length of the day at the place of observation. When the sun is below the horizon the air loses heat by radiation, and still more, in the strata near the surface, by convection to surfaces cooled by radiation.

It was until lately believed that the experiments of Herschel and Pouillet had given an approximate measure of the absolute intensity of solar radiation, and that the proportion absorbed by the atmosphere at the sea-level at a vertical incidence might be estimated at about one-fourth of the whole. It is not too much to say that the recent researches of Mr. Langley, especially those detailed in his *Report of the Mount Whitney expedition*,* have completely revolutionized this department of physics. It now appears that the true value of the solar constant is not much less than twice as great as the previous estimate, and that rather more than one-third is absorbed by the atmosphere before reaching the sea-level. Mr. Langley has further proved that the absorptive action of the atmosphere varies with the wave-length of the rays, and that, omitting the "cold bands" which correspond to the dark bands in the visible spectrum, it diminishes as the wave-length increases. It further appears highly probable that the larger part of the absorptive action of the atmosphere is due to the aqueous vapour, the carbonic-acid gas, and the minute floating particles of solid matter, which are present in variable proportions. Allowing for the probable extension of our knowledge by further research, it is yet evident that, even if we had not to take into account the further elements of the problem next to be specified, the distribution of heat in the atmosphere, as dependent on solar radiation, is a question of extreme complexity.

The action of winds has an important effect in modifying the temperature of the air. It is not possible to draw a distinct line between the great air-currents, which affect large areas, and slight breezes, depending on local causes, and limited to the lower strata of the atmosphere ; but in relation to the present subject it is necessary to distinguish between them. There is a general circulation in the aërial envelope covering the earth, caused by unequal heating of different parts of the surface.

---

* Published by the War Department, United States Army, *Professional Papers of the Signal Service*, No. xv.

Heated air rises in the equatorial zone, and its place is filled by currents from the temperate and subtropical zones. The heated air from the equator flows at first as an upper current towards the poles, but as it gradually loses its high temperature, it becomes mixed with the currents setting from the poles towards the equator, causing the atmospheric disturbances and variable winds characteristic of the cooler temperate zones. As a rule, bodies of air of different temperatures do not very quickly mix, but tend to arrange themselves in layers or strata in which masses of unequal temperature are superposed. It is obvious that in such a condition, where a layer of colder air lies between two having a higher temperature, the whole cannot be in a state of equilibrium. But in nature we constantly find that equilibrium is never attained. There is a continual tendency towards equilibrium, along with fresh disturbances which alter the conditions.

As Professor Stokes remarks in a letter on this subject with which he favoured me, " to know the temperature of the successive strata as we ascend in a balloon, we should know the biographies of the different strata." Those which are now superposed may have been hundreds of miles apart twenty-four hours before. It follows that without a knowledge of the course and velocity of the higher currents existing in the atmosphere, we cannot expect to learn the vertical distribution of temperature.

Apart from the effects of the great movements of the air, there is another effect of air-currents to be considered, which tends especially to modify the temperature found at or near the earth's surface. The heating of the surface by day, and the cooling by night, determine the existence of local currents of ascending or descending air. In rising, the air encounters diminished pressure, and therefore expands, and in so doing overcomes resistance. The molecular work involved in dilatation is performed at the expense of the other form of molecular work which we call heat. In other words, the air in ascending loses heat. It is found that the amount of decrement of temperature due to the ascent of a body of air is nearly exactly $1°$ C. for 100 metres. As a general rule, ascending currents arise from the surfaces exposed to the sun during the day, and must largely contribute to produce the rapid decrement of heat which is found in the lower strata near the surface, as compared with the rate of change in the higher regions; but it will be obvious that the amount of effect produced by this cause is subject to continual variation from changes in local conditions. The nature of the soil, the extent and character of the vegetation, the form of the surface, are all elements which modify the amount of disturbance in the equilibrium of the surrounding atmosphere. As above remarked, in discussing the climate of Western Peru, prevailing winds which impinge upon a range of mountains may indirectly affect the temperature of the higher region by mechanically forcing masses of air to rise along the slopes, and ultimately, by expansion, to be cooled much below the temperature which they possessed when they originally flowed against the slopes.

One of the most important agencies affecting the distribution of temperature in the atmosphere arises from the presence of aqueous vapour. In its invisible condition it affects the absorptive power of the air on the solar rays, and, when condensed in the form of cloud, it acts as a screen, intercepting most of the calorific rays which would otherwise reach the earth. But it is especially through the large amount of heat consumed in converting water into vapour, and set free when vapour returns to the fluid state, that the temperature of the air is largely modified. When we consider that in converting a given volume—say, one cubic metre—of water into vapour, enough heat is consumed to lower about 1,650,000 cubic metres of air by $1°$ C. in temperature, and that the same amount of heat is liberated when the vapour so produced returns to the liquid state, we perceive how powerfully the ordinary processes of evaporation and condensation must affect the temperature of the air.

It is needless to analyze further the several agencies which, sometimes co-operating, and sometimes in mutual opposition, determine the vertical distribution of temperature in the atmosphere. It is but too obvious that no approach to uniformity can be expected, and it might even be anticipated that any approximation to a regular law of distribution that should be found under one set of conditions—as, for instance, in serene weather by day—would be altogether inapplicable under different conditions, such as exist in stormy weather, or by night.

The need for practical application of some empirical rule, or law, of vertical distribution has made it necessary to appeal to the results of observation, and for this object the only existing materials are to be found in the records of balloon ascents, and in the observations made on high mountains. In balloon ascents the temperature at any considerable height is free from the disturbances caused by the vicinity of the earth's surface, and the results might be expected to contribute to the more accurate determination of the amount of atmospheric refraction. For the measurement of heights by the barometer, it would appear safer to rely on such information as may be gleaned from mountain observations.

Of balloon ascents by far the most important are those achieved in 1862 by Messrs. Glaisher and Coxwell, to which I have referred in a preceding page. Mr. Glaisher has given in his report a full record of the actual observations made in the course of his eight ascents, and has explained the processes by which he constructed the successive tables, from which he deduced as the final result a continuous decline (unbroken save in a single instance) in the rate of decrement of temperature found in passing through each successive zone of 1000 feet, in ascending from the sea-level to a height of 29,000 English feet.

I am not aware that the processes employed by Mr. Glaisher in obtaining these results have ever been subjected to such close scrutiny as their importance demands, and as I have found on careful examination that his results are not borne out by the actual observations, I am forced to express my dissent from his conclusions. The admiration due to the courage, skill, and perseverance displayed by Mr. Glaisher throughout these memorable ascents will not be lessened if we should find it necessary to modify the inferences which he has drawn from them.

The full discussion of Mr. Glaisher's observations involves an inconvenient amount of detail, and such readers as may be disposed to enter more fully into the subject I must refer to an article in the *London, Edinburgh, and Dublin Philosophical Magazine.*

The general conclusions to which I have arrived from the observations made under a clear or partially clear sky is, that the average results show a rapid fall of temperature in the zone extending to about 5000 feet, or 1500 metres, above the earth's surface, and that, within that limit, the rate of fall diminishes as the height increases. Above the height specified the observations prove that in each ascent the balloon passed through successive strata of air whose temperature varied in a completely irregular manner, the fall of temperature being sometimes very rapid for an ascent of a few hundred feet, and sometimes very slight in a much longer interval. In each of the higher ascents we even find instances in which the thermometer rose in ascending from a lower to a higher station, reversing the ordinary progression. These alternations occurred at various heights from 5000 to 25,000 or 26,000 feet above the sea-level.* It seems to me very doubtful whether any safe conclusions can be drawn from averages deduced from separate series of observations so discordant, but, in any case, I may confidently assert that the results of actual observations do not bear out the conclusions deduced by Mr. Glaisher.

I desire further to point out that these balloon ascents were all executed by day, in summer, and in weather as serene as can ordinarily be found in our climate. If they did authorize us to derive from them an empirical law regulating the vertical distribution of temperature, this might, at the best, serve to approximate to the true amount of atmospheric refraction found by day in geodetical observations, but would be no guide to the conditions obtaining by night, which are those important to the astronomer.

Mr. Glaisher has not failed to notice the great difference shown by the observations made when the sky was overclouded as compared with those under a clear or partially clear sky, and has given a table showing that the mean results up to a height of 4000 feet above the sea show a nearly uniform decline of 1° Fahr. for each 244 feet at ascent. The numerical results of observations made under, or amidst, cloud appear to me of no practical value, as they depend upon conditions which are subject to constant variation.

If it be true that observations in balloon ascents, which are free from the disturbances caused by the vicinity of the earth's

---

* Air nearly saturated with vapour is lighter than air relatively dry; and hence it may happen that, when a current of moist air meets one relatively dry, it will flow over the latter if they are nearly at the same temperature, but if the drier current be much warmer, it may flow beneath it.

surface, have hitherto failed to lead to any general results indicating a normal rate of decrease of temperature with increasing elevation, it could scarcely be hoped that observations on mountains should contribute farther to enlighten us. From what has been already said, it is apparent that the fact that the place of observation is close to the surface causes disturbances the nature and amount of which must vary with each particular spot, and with the season and the condition of the atmosphere at the moment of observation.

The intensity of solar radiation increases rapidly with increasing elevation,* so that when the sky is clear surfaces exposed to the sun are heated much above the normal temperature. Owing to its slight absorptive power the free atmosphere is little affected; but the strata nearest the surface are heated by convection, while a contrary effect follows when the surface is no longer exposed to the sun, and radiates freely to the sky.

The air in mountain countries is rarely at rest. Even when there is no sensible breeze, the unequal heating of the surface causes ascending and descending currents, which lose or gain heat by expansion or contraction. More commonly winds are experienced which, by impinging on the inclined surfaces, force bodies of air to higher elevations, and thereby directly cause a fall of temperature.

All these causes of disturbance are complicated by the action of aqueous vapour, which, in most mountain countries, is supplied from the surface, as well as borne upwards by ascending currents. Besides the effect of raising the temperature where condensation takes place, and lowering it where clouds are dissolved in strata of dry air, the amount of aqueous vapour present at a given place affects the intensity of solar radiation, and the consequent amount of heating of the surface.

In spite of these obstacles to the attainment of accurate numerical results from which to infer the distribution of temperature in the atmosphere, we are yet, for the larger part of the earth, forced to rely on mountain observations as the only

---

* On this subject see *Handbuch der Klimatologie*, by Julius Hann, pp. 141, *et seq*. See also Tables I. and II. in a report on thermometric observations in the Alps, by J. Ball, in *Reports of the British Association for the Advancement of Science* for 1862, pp. 366-368.

available source from which any indications of a law of distribution can be gleaned. Balloon observations have hitherto, so far as I know, been confined to a few places in Europe; and, even if the results were more conclusive than they have hitherto been, we should not be entitled to infer that they held good for all parts of the earth. In countries where the course of the seasons is more uniform, and the direction and force of the winds less inconstant, it might be expected that the distribution of temperature would exhibit some nearer approach to uniformity; and the possibility of making observations at mountain stations by night might enable us to form some conjecture as to a condition of the atmosphere very different from that which obtains when the influence of the sun is present.

It cannot be said that the observations hitherto made on mountains have done as much as they might do, if properly conducted, to contribute to our knowledge; but a few leading facts may be derived from them, and it is worth while to point them out.

The most important of these is, perhaps, the influence of plateaux of elevated land in raising the temperature of the adjacent air. This is established by observation in all parts of the world, and it would appear that the rapid fall of temperature in the strata near the surface which is found at or near the level of the sea, is equally marked when we ascend from a plateau to an isolated summit. Both these conclusions, however, apply only to observations made in the summer of temperate regions, or in the warmer parts of the earth. Apart from this effect of a relatively heated surface which appears to extend above the surface to a height of about 1500 metres, or, in round numbers, 5000 English feet, mountain observations give but slight confirmation to the belief that the rate of decrease of temperature, in normal conditions of the atmosphere, diminishes as the elevation increases.

In endeavouring to use the available materials one difficulty arises from the fact that, in comparing the temperature of the upper with the lower stations, observers have rarely been supplied with simultaneous observations at the lower station, or that, when these have been available, the distance has been so great that the results throw little light on the probable condition of a vertical column of air near the higher station. In parts of the world where the daily range of temperature near the coast is very slight, we may with small risk of error use the mean temperature of the season at the lower station as the element of comparison, and, in places near the equator, the mean annual temperature. For this reason, observations in the Andes of Ecuador, Peru, and Bolivia present great advantages, and I think it may be useful to discuss the results so far as they are now available.

It is scarcely necessary to examine critically the results of the earlier explorations. Humboldt has given in the "Recueil des Observations Astronomiques," etc., and in the "Memoires de la Société d'Arcueil," vol. iii. p. 579, and elsewhere, the observations made by himself in Mexico, Colombia, and Peru, and also those of Caldas and Boussingault, and has derived from them a table which, with more or less modification, has been adopted in many physical treatises. It exhibits the mean differences of temperature found in successive zones differing in height by 500 toises, the interval corresponding to 974·6 metres, or very nearly 3000 English feet.

Height in toises.	Mean temperature.	Number of metres corresponding to a fall of 1° C. from the sea-level.	Number of metres corresponding to a fall of 1° C. between successive zones of 500 toises.
Sea-level	27·5	—	—
500	21·8	171	171
1000	18·4	216	287
1500	14·3	221	238
2000	7·0	190	133
2500	1·5	187	177

The first remark to be made about this table is that the observations on which it is founded are not properly comparable, being partly single observations made during an ascent, and partly the mean of numerous observations made at certain places, such as Mexico, Quito, etc. It may further be remarked that many of the heights determined by Humboldt have been considerably modified by the results obtained by more recent travellers, and cannot now be regarded as correct. The influence of plateaux is, however, very apparent, as nearly all the observa-

APPENDIX.   381

tions from which the estimated temperatures for 1000 and 1500 toises were derived were made at places situated on open elevated valleys or plateaux. At the utmost, the results can be regarded merely as rough approximations to the truth.

By far the most important available observations in the Andes are those of Mr. Whymper, made during his remarkable explorations in 1880; but, unfortunately, the details have not yet been given to the world, and, in endeavouring to make use of them, I have been forced to content myself with the brief summary published in the *Proceedings of the Royal Geographical Society* for 1881. Mr. Whymper was able to secure a register of the temperatures observed at Guayaquil during his stay in Ecuador, which will doubtless be published along with the record of his own observations; but it does not appear that he was able to obtain observations at Quito during his ascents to the higher peaks; and it seems that, in comparing the temperatures for the purpose of reducing his barometrical observations, he was forced to assume for Quito a mean temperature of 57°9 Fahr., or 14°4 C., obtained from a series of thermometric observations made during his stay at that place. There is reason to believe that the daily range of the thermometer at Quito is very moderate; and at the equator the differences of season are comparatively slight; nevertheless, the absence of simultaneous observations at that place diminishes the value of the results shown in the following table, in which Mr. Whymper's results are reduced to metrical measure.

I have adopted the heights determined by Mr. Whymper as those deserving most confidence. They agree very well with those published by MM. Reiss and Stubel, so that the limits of error from this cause are inconsiderable. I have also adopted the height assigned to Quito—9350 feet, or 2848 metres. Where Mr. Whymper remained long enough on any summit to observe notable variations in the reading of the thermometer, I have taken the mean of the observed temperatures; but I have entered separately the results of the ascents of Chimborazo, one being made in January, the other in July, and in a separate line I have entered the mean results of the two.

In the following table I have entered in the first column the names of the peaks ascended by Mr. Whymper; in the second, the height of each as given by him; in the third, the observed temperature in degrees Centigrade; in the fourth, the difference between the observed temperature and 27° C.—that assumed for Guayaquil; in the fifth, the average number of metres corresponding to a fall of 1° C. in rising from the sea-level to the higher station; in the sixth, the difference between the observed temperatures and that assumed for Quito—14·4°; and in the seventh, the average number of metres corresponding to a fall of 1° C. in rising from Quito to the higher station. It is obvious that the more rapid the fall the less will be the number in columns 5 and 7.

	Name of mountain.	Height above sea-level.	Observed temperature.	Difference of temperature at sea-level.	Average number of metres for fall of 1° C. from sea-level.	Difference of temperature at Quito.	Average number of metres for fall of 1° C. from Quito.
1	Chimborazo (Jan.)	6253	− 6·1	33·1	189	20·5	166
2	Chimborazo (July)	6253	− 8·06	35·06	178	22·46	151
3	Mean of (1) and (2)	6253	− 7·08	34·08	183·5	21·48	158·5
4	Cotopaxi	5959	− 8·4	35·4	168	22·8	136·5
5	Antisana	5870	+11·1	15·9	369	3·3	916
6	Cayambe	5852	+ 2·5	24·5	239	11·9	252
7	Cahihuairazo	5035	+ 4·44	22·56	223	9·96	220
8	Cotocachi	4965	+ 2·2	24·8	202	12·2	173·5
9	Pichincha	4851	+ 7·77	19·23	255	6·63	302
10	Corazon	4837	+ 4·44	22·56	214	9·96	200
11	Sara Urcu	4718	+10·0	17·00	284	4·4	425

It will at once be seen that the temperatures observed on Antisana, Pichincha, and Sara Urcu were altogether exceptional, probably due to rapid condensation of vapour; and these may best be excluded from any discussion of the general results. The temperatures noted in the second ascent of Chimborazo were probably below the mean, or at least below the mean for the hours at which most of the other observations were made. But, as opinions may differ on that point, I have also given below the results of comparison with the mean for the two ascents of Chimborazo. For a similar reason I regard the figures for Cotopaxi, where Mr. Whymper remained for twenty-six hours on the summit, as giving too low a temperature, while that observed on Cayambe is certainly too high. The mean result for these two summits is probably a near approximation to the average for that height.

In attempting to draw conclusions from the above table, we must first remark that, in consequence of its position on a plateau, the temperature of Quito is considerably higher than it would probably be if the higher peaks descended with an uniform slope to the sea-level. The difference between the means for that place and Guayaquil is only 12·6° C.; whereas, on the supposition of an uniform decrease in ascending from the sea-level, it should be 14·2°, and still greater if we supposed that the rate of fall of temperature gradually diminishes as the elevation increases. Omitting altogether the results for numbers 5, 9, and 11 in the above table, we perceive that the observations fall into three groups: (1) those for Chimborazo, at 6253 metres; (2) those for Cotopaxi and Cayambe, with a mean height of 5905 metres; (3) those for Cahihuairazo, Cotocachi, and Corazon, whose mean height is 4950 metres. To these it may be well to compare the mean of the results for the entire series, and also the rate of decrease between the sea-level and Quito. I shall designate observations included hereunder by numbers corresponding to the lines in the preceding table. The number of metres of ascent corresponding to a fall of 1° C. gives the most convenient measure of the rate of decrease.

	Mean height.	Difference of temperature at sea-level.	Metres for fall of 1° C. from sea-level.	Difference of temperature at Quito.	Metres for fall of 1° C. from Quito.
Quito	2848	12·6	226	0	0
Mean of 1, 4, 6, 7, 8, and 10	5483·5	27·19	201·5	14·59	180·6
,, 3, 4, 6, 7, 8, and 10	5483·5	27·35	200·5	14·75	178·7
,, 7, 8, and 10 ...	4946	23·37	212	10·77	195
,, 4 and 6	5905	29·95	197	17·35	176

We see from this table that, in ascending from the coast to the highest peaks of Ecuador, the average fall of the thermometer was, in round numbers, 1° C. for every 200 metres of ascent, while in ascending from the sea-level to the plateau of Quito the fall was proportionately less, being at the rate of 1° C. for 226 metres. On the other hand, the fall of temperature was more rapid in ascending from Quito to the higher peaks. On an average of all the ascents, we may reckon the rate of 1° for 180 metres. But it is remarkable that, taking the average of the three peaks which rise about 2000 metres above the level of Quito, the temperature fell only at the rate of 1° for 195 metres, while in ascending to peaks higher by nearly 1000 metres, the rate of fall was 1° for 176 metres, and if we take the still higher summit of Chimborazo we may reckon the rate of fall at about 1° for 160 metres.

The apparent increase in the rate of decline of temperature in the higher region is still more clearly shown if we compare the mean of the three peaks whose average height is 4946 metres, with that of the two whose average height is 5905. For a difference in the mean height of 959 metres, we find an average fall of 6·58° C., or a fall of 1° for 145 metres. Taking the first ascent of Chimborazo as giving the most probable results, we find that between this peak and the mean of the three lower summits, with a difference in height of 1307 metres, the difference of temperature is 9·73°, or a fall of 1° for 134 metres. Again, comparing Chimborazo with the mean of Cotopaxi and Cayambe, we find, for a difference of height of 348 metres, a difference of temperature of 3·15°, or a fall of 1° for 110 metres.

I am fully aware that these observations are not numerous enough to lead to any safe general conclusions; the comparatively high temperatures found at the height of about 5000 metres may be due to exceptional local conditions, such, for instance, as the ordinary formation of clouds at about that level; but, so far as they go, the observations tend to negative the supposition that in the tropics the rate of decrease of temperature diminishes as we ascend to the higher regions of the atmosphere.

MM. Reiss and Stübel made numerous observations in the Andes of Ecuador and Peru, during a prolonged visit to that region. Lists of heights obtained by reduction from their observations have appeared in various German scientific periodicals, and more fully in the *American Journal of Science*, vol. ii. pp. 268, 269 ; but, so far as I can ascertain, the record of their observations of the barometer and thermometer has never been given to the world.

In "Copernicus," vol. iii. p. 193, *et seq.*, Mr. Ralph Copeland has published a summary of the results of a series of meteorological observations made by him at various stations on the line of railway connecting Mollendo on the Pacific coast with Puno in Bolivia, near the lake of Titicaca, and also at La Paz and at Tacna. Two series of observations were made at Vincocaya, the summit station of the railway, 4377 metres above the sea. All the other stations are either on elevated plateaux, or on open slopes inclining gently towards the coast. The temperatures are partly derived from numerous observations and partly by taking the mean of the maxima and minima, with corrections for each station, the reasons for which are assigned by Mr. Copeland. In most of these I am inclined to concur, but there are two from which I am forced to dissent. In reducing Mr. Copeland's tables to metrical measure, I have therefore ventured to make some corrections, which do not, however, much alter the results.

I give below the heights above the sea, in metres, with the corrected mean temperature for each place, and the dates for each set of observations.

Places.	Latitude.	Height.	Dates of observation.	Mean temperature, corrected.
Mollendo	17° 2' 54"	20	July 2	16·7° C.
Tacna	18° 1' 21"	560	July 7–10	14·2°
Arequipa Hotel	16° 25' 20"	2346	Feb. 2–8	16·2°
Arequipa railway station	—	2300	June 29–30	9·0°
Vincocaya, I.	15° 53' 56"	4377	Feb. 28–March 4	2·83°
———, II.	—	—	June 6–27	−2·2°
Puno, I.	15° 50' 2"	3840	March 20–April 4	9·2°
———, II.	—	—	April 15–June 2	7·8°
La Paz	16° 27' 0"	3645	Feb. 12–25	10·7°

Without entering into minute details, or discussing the small corrections for changes in the sun's declination to be allowed for latitude and for the dates of observation, we perceive that on the western slope of the Cordillera the rate of decrease of temperature in this region is much below the ordinary average. Estimating the mean temperature of Mollendo at 22° at the beginning of February, we find between Mollendo and Arequipa a difference of 5·8° C., or a fall in summer of 1° for an ascent of 401 metres; while in mid-winter we obtain a difference of 7·7°, showing that an ascent of 364 metres is necessary to cause a fall of 1°. This abnormal condition is, no doubt, mainly due to the exceptionally low temperature of the coast-zone. Between Arequipa and Vincocaya we may reckon the fall of temperature on the 1st of March at 14·2° for an ascent of 2031 metres, giving the proportion of 1° to 143 metres; but in winter the decrease is less rapid, as we have at the end of June a difference of about 11·5° for an ascent of 2077 metres, or about 181 metres for a fall of 1°.

A remarkable contrast is shown when we compare the temperature at Vincocaya with that of places on the plateau surrounding the great lake of Titicaca. From Mr. Copeland's observations we may estimate the mean annual temperature of Vincocaya at 1° C., that of Puno at 8·5°, and that of La Paz at 8·8°. These figures would give a mean difference of 7·5° for a difference in height of 537 metres between Vincocaya and Puno, or a decrease of 1° for 72 metres. Between Vincocaya and La Paz we have a difference of 7·8° for a difference in height of 732 metres, or a fall of 1° for 94 metres. The mean of the two comparisons gives a fall of 1° for 83 metres, or about twice as rapid a change as the average of the comparison between Arequipa and Vincocaya. I am not disposed to attribute this remarkable difference of atmospheric conditions exclusively to the influence of plateaux in raising the mean temperature.

In my own slight experience in the Peruvian Andes, in ascending from Chicla, at about 3700 metres, to Casapalta, at about 4200 metres, I observed so complete and rapid a change in the character and aspect of the vegetation as to satisfy me that the difference in the annual mean temperature must be even greater than that observed by Mr. Copeland for a somewhat greater difference of height between Vincocaya and Puno. It may be that, in this comparatively dry region of the Andes, the higher stations receive more frequent, though not copious, falls of rain or snow, the evaporation of which maintains a constant low temperature in the surface and the surrounding air.

In comparing observations in Peru, Bolivia, or Chili with those made in the Andes of Ecuador, it must not be forgotten that the climatal conditions are essentially different. Owing to the fact that in the latter the range of the Andes is much narrower, and on one side the main valleys descend in a nearly due easterly direction, the hot, vapour-laden, easterly winds reach the plateaux still charged with moisture, and at all seasons rain is frequent and abundant. Farther south, the winds from the Atlantic have deposited the greater part of their moisture before they arrive at the western side of the main range, and the annual rainfall must be comparatively trifling.

I have sought in vain in the records of mountain observations in other parts of the world for materials from which any probable inference may be drawn as to a law regulating the ratio of decrease of temperature with increasing height above the sea-level. There is reason to admit that isolated peaks of no great height show a more rapid decrease as compared with the plain than do considerable mountain masses. Of mountains exceeding the height of 3000 metres in the tropics, the most rapid rate of decrease is that recorded for Pangerango in Java, being 1° for 178·5 metres.

The greater mountain masses in or near the tropics show nearly the same rate of decrement, by comparison with the sea-level, that I have been led to infer from the observations in Ecuador. The average rate for the Himalayas is about 1° for 194 metres of ascent, and for the less lofty peaks of Mexico Humboldt's observations show a decrease of 1° for 188 metres. The great irregularities due to local conditions make it impossible to derive any positive conclusions as to the comparative rate of decrease in successive zones of elevation.

In Europe and North America comparisons between the temperatures at mountain summits and the sea-level give rates of decrease varying between 1° for 160 metres, and 1° for 170 metres; but it must be remarked that the averages are mainly founded on observations made in summer, and it is certain that the rate of decrease is much slower in winter. Where the difference of height is not very great, it not uncommonly happens that in winter the phenomenon is reversed, and that we experience an increase of temperature in ascending above the plain. The same result on a small scale may often be remarked on clear cold nights, when the temperature rises for a distance of some hundred feet in ascending isolated eminences, the effect being due to the cooling effect of radiation from the surface.

It seems most probable that in the winter of the temperate and polar zones the distribution of temperature in the atmosphere is subject to conditions widely different from those prevailing in summer; and, if that be true, we should have intermediate conditions in the spring and autumn; so that even if we could arrive at comparatively accurate results for one season of the year, these would not be applicable at other periods.

The general result to which I have arrived is that to ascertain the distribution of temperature in the atmosphere in successive zones of elevation is a problem of extreme complexity, towards which the existing materials do not furnish even an approximate solution. I hold, however, that it ought to be possible to obtain much more definite knowledge than we now possess by means of properly conducted observations in various parts of the world.

Foremost of these I would suggest the importance of well-conducted balloon ascents within the tropics. In selecting stations for such ascents we are somewhat restricted by local considerations, especially the extension of forests in many regions, such as the greater part of tropical Brazil. In British India there would be no difficulty in selecting suitable stations, and there would be additional value in comparing the results obtained from ascents in Bengal, and in the very different climate of the North-west Provinces. Elsewhere in the tropics we might expect valuable results from ascents in Queensland, and from the *llanos* of Venezuela. It seems not impossible that, with a considerably smaller outlay, useful results may hereafter be obtained by means of improved self-recording instruments sent up in captive balloons; but in most countries such a record would be liable to interruption owing to storms.

The next desideratum is to obtain for a series of years simultaneous observations at successive stations, at vertical intervals of 500 or 600 metres, situated on the flanks and at the summits of high mountains to be chosen for the purpose. Some of these might with advantage be chosen on islands, and among these the following may be suggested :—the Peak of Teneriffe, Mauna Kea in the Sandwich Islands, Fusiyama in Japan, the Piton de Neige in the island of Réunion, and Etna in Sicily. It would add much to the value of these observations if in each case there were a double series of stations, one series being on the windward, the other on the leeward side of the mountain. It would also be important to obtain observations at similar series of stations in continental regions, removed from the immediate influence of the sea. Pike's Peak in Colorado, which already possesses an observing station at the summit, and Mount Whitney in California, which Mr. Langley has selected as eminently suited for an observatory, both offer many advantages for the desired purpose. Another desirable station might easily be found in the Caucasus, or in Armenia, and one or more could be selected on the southern declivity of the Himalayes. In South America, where railways have been carried to such great heights, it may be hoped that regular observations may at some future time be secured at the successive railway stations. It would be worthy of the enlightened governments of Chili and Argentaria to make a commencement, by providing for such a series being obtained at the stations on the railway now in course of construction over the Uspallata Pass.

For the realization of most of these desires, as well as many others affecting the progress of human knowledge, and the general welfare of our race, we must be content to await the advent of a happier era, when the fruits of industry, and the efforts of rulers, shall no longer be mainly devoted to the maintenance and development of the arts of destruction.

While awaiting such additional knowledge as may hereafter be obtained, it is necessary in the mean time to form some provisional hypothesis on which to base the formulæ for determining the difference of heights of two stations, by barometric observations, and for ascertaining the amount of atmospheric refraction; and the subject might with advantage be discussed at a congress of scientific men. I have no authority to decide on a question of such difficulty, nor do I pretend to be thoroughly versed in the somewhat voluminous literature of the subject. I may remark, however, that in one of the fullest and most elaborate works by recent writers, Dr. Rühlmann.[*] has proposed a formula for the reduction of barometric observations which implicitly assumes that the rate of decrement of temperature in ascending mountains is uniform, inasmuch as he takes the mean of the temperatures observed at the higher and lower stations as the value of the mean temperature of the column of air between the two stations. It would appear that his adoption of the hypothesis of an uniform rate of decrease is merely due to the apparent impossibility of discovering a more satisfactory hypothesis. Following on a line of inquiry first suggested by the late M. Plantamour and M. Charles Martins, Dr. Rühlmann has analyzed a series of two-hourly observations of temperature made during six years at the hospice of the Great St. Bernard and at the Geneva Observatory. Treating the mean temperature of the column of air between the levels of those places as the unknown quantity, and neglecting, as unimportant, the corrections for the tension of aqueous vapour and for gravity, he has deduced the "true temperature," as he styles it, of the intermediate column from the equation of condition between the pressures, the heights, and the temperatures of the two stations, for the average of the two-hourly periods of observation for each month. He has shown that, while on the average of the entire year the mean "true temperature" of the intermediate column of air agrees pretty well with the mean of the yearly observations at the two extreme stations, the means for the separate hours and those for the separate months usually differ widely from the so-called "true temperatures" for the corresponding periods.

From this investigation Dr. Rühlmann has shown that during the warm hours of the day, and the summer months, the "true mean temperature" is lower than the mean of the observed temperatures at the two extreme stations, while at night, and during winter, it exceeds that mean to a rather greater extent. It may be objected that the cause of the apparent discrepancy lies in the fact that, in thermometric observations, we obtain, not the true temperature of the surrounding air, but that of the thermometer, and that, however carefully screened, the thermometer cannot be completely freed from the effects of radiation

---

[*] See "Die Barometrischen Höhenmessungen und ihre Bedeutung für die Physik der Atmosphäre," Leipzig, 1870, by R. Rühlmann.

to and from surrounding objects. This remark applies especially to the observations at the St. Bernard, which lies at a considerable distance from Geneva, and where the temperature is unduly depressed by surrounding masses of snow. I do not, however, attach much importance to these sources of error; and I have no doubt that under the most favourable conditions the discrepancy shown by Rühlmann will be found to a greater or less extent, but I differ from that writer in the inference that he has drawn from the facts.

If I have not misunderstood his remarks, Dr. Rühlmann concludes that the true temperature of the successive strata of air in the zone between the base and the summit of a mountain is but slightly affected by the diurnal changes that are exhibited in the range of the thermometer, and to a moderate extent only by the changes of season as shown by the range of the monthly means. He has not adverted to the fact that the differences disclosed in his tables may be the result of changes in the rate of decrement of temperature in ascending from the lower to the higher station. He shows that, on the mean of the July observations, the mean temperature of the air between the levels of Geneva and the St. Bernard is lower than the mean difference of the temperatures observed at those places by 1·57° C. But this is not inconsistent with the supposition that the thermometers have recorded the true air temperature at each station, but that the rate of decrement of temperature in ascending, at that season, diminishes rapidly in the successive vertical zones. In the same manner the fact that the true mean temperature in January is higher than the mean of the observed thermometers by 1·83° C., might be accounted for by supposing that in winter the rate of decrement is smaller in the lower strata, and increases in ascending above the surface. It is equally true that, in both cases, the facts may be consistent with such an irregular distribution of the atmosphere in successive layers, or strata, of very unequal temperature as was apparent in most of Mr. Glaisher's balloon ascents. What is completely proved is that it is only under exceptional conditions that the hypothesis of an uniform rate of decrement of temperature, directly proportional to height above the sea-level, is approximately correct for observations in the temperate zone, where there is a considerable diurnal and annual range of the thermometer.

My own impression, as the result of such study as I have been able to give to the subject, is that, in the present state of our knowledge, the reduction of barometric observations for the height of mountains made by day, and in summer, in temperate latitudes, may best be effected by the formula proposed by M. de St. Robert; while for observations made at other seasons, and in the tropics, I should prefer the formula proposed by Mr. Rühlmann.

Before closing these remarks, I may refer to an ingenious suggestion made by M. de St. Robert in a paper published in the journal *Les Mondes* in Paris, in 1864, the substance of which is to be found in the *Atti dell' Academia delle Scienze di Torino* for 1866, p. 193. Impressed with the difficulty of approximating in practice to a correct knowledge of the distribution of temperature in the air between the summit of a mountain and a lower station, the author sought to escape from it by seeking a phenomenon, susceptible of observation, which should give a direct measure of the mean density of the air in the space between the two stations. He pointed out that the velocity of sound supplies such a measure, and that, given the barometric pressures at the higher and lower stations, the angle of elevation of the former, measured by a theodolite and corrected for refraction, and the exact time required for sound to traverse the interval between them, the height is given with a near approximation to accuracy by a simple formula. The error arising from air currents, which increase or diminish the velocity of transmission, would be readily eliminated by discharging a fire-arm simultaneously at both stations, observing the interval between the light reaching the eye and the report becoming audible, and taking the mean of the intervals observed at both stations.

M. de St. Robert does not disguise the practical difficulty of measuring the time interval with the requisite accuracy, but he thinks that it may be obtained within a fifth of a second. The error in the result is inversely proportionate to the time required to traverse the distance, and where the stations are as distant as is compatible with the sound being audible, its amount for an error of a fifth of a second is inconsiderable.

This suggestion has not received the attention which it seems to deserve. It possesses the advantage that the observations may readily be repeated with little trouble or cost, and that the risk of error may be much diminished by taking the mean of the observed intervals of time. A comparison between observations between stations whose height is known, made under different conditions, by day and night, and in different states of weather, might, I think, contribute to diminish our ignorance as to the variable conditions of the atmosphere at different heights above the surface.

APPENDIX B.

REMARKS ON MR. CROLL'S THEORY OF SECULAR CHANGES OF THE EARTH'S CLIMATE.

MOST scientific readers are familiar with the theory respecting the influence of changes in the eccentricity of the earth's orbit on the climate of the globe, which has been sustained with remarkable ability by Mr. James Croll. The views originally advanced in various scientific periodicals were presented to the public in a connected form in the volume entitled "Climate and Time," wherein the author has brought a wide knowledge of the principles of physics, and of the whole field of geological science, to the support of his theory. Even those who have not given especial attention to the subject are also acquainted with the conclusions which Sir Charles Lyell drew from the discussion of Mr. Croll's arguments, and which are contained in the thirteenth chapter of the tenth edition of his "Principles of Geology," and also with the more recent examination of the subject which is to be found in Mr. Alfred Wallace's important work, "Island Life."

I need not say that a theory so important in its bearing on some of the most obscure problems of geology has been discussed, in more or less detail, by many other writers. To most of the objections presented to his theory, Mr. Croll has replied with his usual ability; and I believe that at present the prevailing tendency among geologists is towards a partial acceptance of his views, subject to the limitations assigned by Mr. Wallace.

The latter author holds, in common with Sir Charles Lyell, that geographical causes, arising from the varying distribution of land and sea, have mainly controlled the distribution of temperature over the earth's surface ; but he is disposed to go farther than Lyell in admitting the influence of periods of high eccentricity in causing those great accumulations of snow and ice which were requisite to produce the phenomena of a glacial period, whenever a sufficient area of elevated land in high latitudes coincided with the period of high eccentricity.

It would probably be of little avail, even if I were to undertake the task, that I should attempt any thorough discussion of this vast and difficult problem ; and it would certainly require far more space than can here be given to it. I may, however, venture to make a few remarks upon some points which have not, to the best of my knowledge, been much noticed in the discussion.

In reading Mr. Croll's work, which charmed many an hour during the voyage to and from South America, I found it very difficult to discover any flaw in the chain of close reasoning by which he supports his conclusions. Most of the facts on which he relies are warranted by observation, and have been accepted as well established by writers of the highest authority ; and his inferences as to the results of altered conditions appeared to be in strict conformity with admitted physical principles. Nevertheless, when I reflected on the anomalies which are found at the present time in respect to the climate of many spots in the world, and the complexity of the causes which determine its actual condition, I felt a doubt whether, in his attempt to trace the result of possible changes, Mr. Croll may not have overlooked some of the elements of the problem.

Let me briefly state the leading propositions of Mr. Croll's theory in order to make intelligible the succeeding remarks.

Estimating approximately the mean distance of the earth from the sun at ninety-one and a half millions of miles, and the eccentricity* of the sun's place in the orbit at one and a half million, it follows that at one period of the year, which happens to be about the winter solstice of the northern hemisphere, the earth receives from the sun a quantity of heat greater than that

---

* I use the term "eccentricity" in the popular sense, to express the distance of the focus from the centre of the ellipse.

which reaches it in the opposite part of its orbit, in the proportion of 93^2 to 90^2, or about as 1000 to 936. Midsummer of the southern hemisphere is the season when the earth is nearest to the sun ; the winter of the southern and the summer of the northern hemisphere occur when the earth is farthest from the source of heat. The conclusion seems inevitable—the southern hemisphere must have hotter summers and colder winters than our hemisphere, where the heat of summer is tempered by the greater distance, and the cold of winter mitigated by the comparative nearness, of the sun.

The next point to be considered is the effect of ocean-currents, and especially of the Gulf-stream, in modifying the climatal conditions of some parts of the earth. Following in the track of the late Captain Maury and Principal Forbes, Mr. Croll has especially insisted on the importance of the great current which, issuing from the Gulf of Mexico, and flowing northward between Florida and the Bahamas, extends across the Atlantic towards the western shores of Europe. He calculates that by this current alone an amount of heat equal to that received on the entire surface of the earth in a zone thirty-two miles in breadth on each side of the equator is carried from the tropics to the cooler regions of the northern hemisphere. Mr. Croll has, I think, victoriously replied to several of the objections opposed to this portion of his argument. His estimate of the volume of water transferred by the Gulf-stream from the tropics to the northern part of the Atlantic, which he reckons at the annual amount of about 166,000 cubic miles, is, I think, in no degree exaggerated ; and I also think that he is warranted in estimating the mean initial temperature at about 65° Fahr. I am, however, persuaded that in assuming 40° Fahr. as the temperature to which, on an average, this vast body of water is reduced before it returns to the equatorial zone, Mr. Croll has gone beyond the probable limit. A large part of the stream is diverted eastward about the latitude of the Azores, and is never cooled much below 55° Fahr. before the waters enter the return current on the eastern side of the Atlantic basin ; and I believe that, if we allow the water of the Gulf-stream to undergo an average loss of temperature of 20° Fahr., we shall be more likely to exaggerate than to underrate the amount of cooling.

In insisting on the importance of the Gulf-stream in modifying the climate of Europe and the adjacent parts of the arctic zone, Mr. Croll agrees with many preceding writers ; but, so far as I know, he was the first to suggest that in consequence of the greater persistency of the south-east trade-winds, which ordinarily extend up to, and, at some seasons, even north of, the equator, the warm waters of the Northern Atlantic derive a large share of the heat which is carried to the temperate and arctic zones from the southern hemisphere. Applying the same reasoning to the currents of the Pacific Ocean, Mr. Croll arrives at the general conclusion ("Climate and Time," p. 94) that "the amount of heat transferred from the southern hemisphere to the northern is equal to all the heat falling within fifty-two miles on each side of the equator."

I do not believe that the facts on which Mr. Croll bases this essential portion of his theory are sufficiently established. With regard to the Atlantic, I have expressed in the text (p. 344) an opinion, derived from conversations with practical seamen, that in the Atlantic the trade-winds of the northern are stronger than those of the southern hemisphere. That opinion, I am disposed, on further examination, to regard as incorrect. I believe that the north-east trade-winds often blow with greater force ; but, taking the average of the entire year, I now think there can be no doubt that the south-east trade-winds extend over a wider area in the equatorial zone. However this may be, our knowledge of the currents of the Atlantic does not, I think, authorize us to conclude that the portion of heated water carried from the southern to the northern hemisphere is nearly so large as Mr. Croll has estimated. If the heat of the Gulf-stream were mainly supplied, as Mr. Croll contends, from that source, there should be a marked difference in the volume and temperature of the current, between the season when the north-east trade-winds approach the equator and that in which the south-east trades prevail to the north of the line, for which there is no evidence.

As regards the currents and winds of the Pacific, in spite of one considerable exception, to which I shall further allude, I think that the balance of evidence points to a greater prevalence of the south-east trade-winds, and to the probable transference of some portion of the equatorial waters from the southern to the northern hemisphere.

For the present discussion it is best to accept Mr. Croll's estimate, and to compare the amount of heat which he supposes to be transferred from one hemisphere to the other with the total amount which is received annually from the sun on each hemisphere. For this purpose I have taken the known areas of the torrid, temperate, and frigid zones respectively, and, following Mr. Croll, I have adopted Mr. Meech's estimate of the average amount of heat, per unit of surface, received from the sun in each zone, irrespective of absorption by the atmosphere. To estimate the proportion of heat which actually reaches the surface, I have adopted Pouillet's measure of the proportion of solar radiation cut off at vertical incidence, which is 24 per cent. I assume 28 per cent. to be the average loss in the torrid zone, 50 per cent. in the temperate zone, and 75 per cent. in the frigid zone.* The resulting figures, showing the proportional amount of heat annually received on the surface of each zone, and on the entire hemisphere, are as follows:—

Torrid zone	3370
Temperate zone	2304
Frigid zone	112
Whole hemisphere	5786

Calculating, on the same basis, the amount received on a zone one mile wide at the equator, allowing a loss of 25 per cent. from atmospheric absorption, and multiplying the result by 104, I obtain the number 233·1, or rather more than one twenty-fifth part of the entire heat annually received from the sun by each hemisphere.

To trace the results of such a transfer of heat from one hemisphere to the other, I shall adopt a mode of reasoning, sanctioned by the great authority of Sir John Herschel, to which Mr. Croll frequently resorts. It is by solar heat that the surface of the earth is raised above the temperature of space, which is assumed to be 239 degrees below the zero of Fahrenheit's scale. Adopting Ferrel's estimate, I take the mean temperature of the northern hemisphere at 59·5° Fahr., or 298½ degrees above the temperature of space. To maintain this temperature, it

---

* Viewed in the light of Mr. Langley's recent researches on solar radiation, all these numerical determinations are probably far from the truth; but the errors do not much affect the present argument.

receives one-half of the amount of solar radiation which reaches the earth, and in addition, on Mr. Croll's hypothesis, one twenty-fifth part of that which reaches the southern hemisphere. It follows that the heat available to raise the southern hemisphere above the temperature of space stands to that which is received by the northern hemisphere in the ratio of 24 : 26, and that the mean temperature of the southern hemisphere should be $298\cdot5 \times \frac{12}{13}$, or $275\cdot5°$ above the temperature of space ; so that, in ordinary language, the mean temperature of the southern hemisphere should be $36\cdot5°$ Fahr. If the fact corresponded with this result of theory, it would not be necessary to invoke increased eccentricity of the earth's orbit to account for the extreme cold of one hemisphere, seeing that the actual conditions would suffice to completely alter their relative temperatures.

It occurs to me, however, that, on further consideration, Mr. Croll would reduce his estimate of the volume of heated water transferred from the southern to the northern hemisphere ; but even if that estimate were reduced by one-half, we ought to find in the southern hemisphere a mean temperature of $47\cdot8°$ Fahr., or nearly 12 degrees lower than that of our hemisphere.

We have already seen that, so far as climate depends on the relative position of the earth and the sun, we ought to find in the southern hemisphere climates of a more extreme character, with hotter summers and colder winters, than those to which we are accustomed. If it be true that through the agency of ocean-currents a considerable amount of heat is transferred to the northern hemisphere, that circumstance might serve to account for the fact that the summers of the southern are not generally hotter than those of the northern hemisphere ; but it would, at the same time, tend to aggravate the severity of the southern winters.

At the time of the publication of Mr. Croll's earlier memoirs, there existed a general belief that the southern hemisphere was in fact notably cooler than our portion of the globe, and he naturally referred to the supposed fact as harmonizing with the general conclusions drawn by him from theory. But, imperfect as our knowledge of the southern hemisphere still is, a good deal of information has been obtained of late years. The only stations south of the fiftieth degree of latitude from which we possess continuous observations are those mentioned in the text (p. 273); but we also know with sufficient accuracy the climates of two widely separated islands lying about 50° south; and from these we derive results widely different from those to which we were led by theoretical considerations. The following table gives approximately the mean temperatures, on Fahrenheit's scale, for the year and for the hottest and coldest months of the places referred to in the southern hemisphere, and the means for corresponding latitudes in the northern hemisphere:—

	S. latitude.	Temperature of January.	Temperature of July.	Mean of year.	N. hemisphere. July.	N. hemisphere. January.	N. hemisphere. Yearly mean.
Kerguelen Land	49° 17′	44·3°	35·3°	39·6°	63·3°	22·0°	42·9°
Auckland Island	50° 30′	50·2°	33·6°	44·6°	62·3°	19·0°	41·1°
* Falklands (Stanley) I.	51° 41′	49·6°	36·5°	43·0°	61·6°	17·1°	39·8°
Falklands II.	52° 5′	55·9°	37·4°	47·3°	61·3°	16·4°	39·3°
Falklands, mean of I. and II.		52·7°	37·0°	45·1°	61·5°	16·7°	31·6°
Punta Arenas	53° 25′	51·4°	34·7°	43·0°	60·6°	14·2°	37·7°
† Ushuaia	54° 53′	53·2°	31·8°	41·9°	59·6°	12·0°	36·2°

If we compare the mean results of these five stations with those for corresponding latitudes in the northern hemisphere, we find that the summers are cooler and the winters very much milder, and that in the latitudes between 50° and 55° the mean annual temperature is notably higher. In Kerguelen Land alone the mean annual temperature is lower than the normal for the same latitude north of the equator; but that island is evidently exposed to exceptional conditions.

---

* The observations at Stanley Harbour, which are those adopted by Dr. Hann (*Klimatologie*, p. 697), show temperatures notably lower than those recorded for a place in the islands lying farther south, which are given in the *Zeitschrift der Œsterreichischen Gesellschaft für Meteorologie*, vol. v. p. 369. The mean of the two is probably nearly correct.

† These figures are derived from the tables given in the *Anales de la Oficina Meteorologica Argentina*, by B. Gould, vol. iii. The figures show a considerable amount of annual variation. The monthly means of the six months from February to July, 1879, exceed those of the same period in 1878 by more than 2° Fahr.

The differences between the mean results given above are shown by the following table, in which the signs show the excess or deficiency of the southern as compared with the northern hemisphere :—

Warmest month.	Coldest month.	Annual mean.
− 11·1° Fahr.	+ 18·1° Fahr.	+ 4·2° Fahr.

Dr. Hann has carefully discussed the question as to the comparative mean temperatures of the two hemispheres in a paper published in the proceedings of the Vienna Academy, the substance of which is given in his *Klimatologie*, pp. 89, *et seq.;* and it is difficult to refuse assent to his conclusion that so far as the available evidence goes, it shows that the mean temperature of both hemispheres is equal.

I find, then, that the same train of reasoning by which Mr. Croll has sought to explain the occurrence of glacial periods by changes in the eccentricity of the earth's orbit, and the precession of the equinoxes, leads us to conclusions respecting the climatal condition of the different parts of the earth, at the present amount of eccentricity, which are altogether opposed to the results of observation ; and I am driven to the conclusion that the causes which he has adduced have not the predominant influence which he has attributed to them, and that there must be other agencies to which he has not assigned their due importance, but which are adequate to counteract the efficiency of those which, as observation proves, fail to achieve the effects anticipated from them.

I am far from pretending to be able to analyze completely the complex agencies which, by their mutual action, determine the climate of different parts of the earth, but I may briefly refer to two of them. Foremost of these is the relative distribution of land and sea, for a due appreciation of which we are indebted to the great work of Sir Charles Lyell. It is unnecessary here to discuss how far his view of the probable amount of change in past geological epochs may, in the present state of our knowledge, be subject to limitation. Mr. Wallace, who is the most strenuous supporter of the modern doctrine of the permanence of the present continents and ocean basins, recognizes the theoretical correctness of Lyell's views, and admits that changes of level great enough to cause profound modifications of climate have actually occurred. Notwithstanding recent objections, it appears to me that Darwin's hypothesis as to the subsidence of a great tract in the Southern Pacific is that which best accounts for the existence of the countless coral islands in that region; nor is the probability of a nearly continuous barrier of volcanic islands across the Atlantic to be completely dismissed. That such changes would have largely affected the climate of the earth cannot, I think, be doubted.

If I may venture to express my own view on this difficult subject, I must say that, although it has not been overlooked by the able men who have discussed it, the paramount importance of aqueous vapour as an agent for modifying climate has not yet been fully recognized. Mr. Croll has constantly discussed the phenomena of ocean-currents, as if their chief function were to affect climate by heating or cooling the surrounding air, which is thence diffused over the land surfaces, and he has devoted little attention to the effects of evaporation from the sea, and the subsequent condensation in some other region of the vapour produced. When we remember that as much heat is consumed in the conversion of one cubic mile of water into vapour as would raise the temperature of nearly ninety-seven cubic miles of water by 10° Fahr., we get some measure of the vast power of vapour as a vehicle of heat. Admitting, as I am disposed to do, that 166,000 cubic miles of water are annually conveyed northward by the Gulf-stream, and suffer an average loss of 20° Fahr. before returning to the torrid zone, I must point out that the entire heat requisite to maintain this great volume of water at the higher temperature would be consumed in the conversion of 3433 cubic miles of water into vapour. In point of fact, I believe that more than one-half of the quantity specified is expended in evaporation, and that the cooling of the waters of the Gulf-stream is mainly due to this agency. To follow the vapour thus produced, to ascertain where it is condensed, and where the heat disengaged in the act of condensation becomes available to raise the temperature of the air, is a task which is beyond our present resources; but it is one which must be performed before we can reason with any confidence as to the ultimate distribution of the heat carried by the Gulf-stream or any other ocean-current. Whatever part of the vapour produced by evaporation from the Gulf-stream goes to supply the rainfall of Western Europe, or to form snow in the arctic regions, acts as a vehicle to transfer heat from the tropics to the temperate and frigid zones. But it is more than probable that a large part of the vapour in question is carried back to the torrid zone, and that some of it is even restored to the southern hemisphere. The south-eastern branch of the Gulf-stream flows, at least partially, into the area of the north-east trade-winds. These winds reach the lower region as cold and very dry winds. As they advance towards the equator, and are gradually warmed, their capacity for aqueous vapour constantly increases, and there can be no doubt that in both hemispheres the trade-winds bear with them a large share of the vapour which goes to supply the heavy rainfall of the tropics.

In the Pacific region we have direct evidence to this effect, in the fact that in Hawaii, and elsewhere, the side of the islands exposed to the trade-winds is that of heavy rainfall, and is generally covered with forest. No sufficient data exist for estimating the amount of vapour thus carried back to the tropics from high latitudes on both sides of the equator, nor the amount of heat set free by its condensation; but we may form some conception of its probable amount by considering that at the moderate estimate of a mean annual rainfall of seventy-two inches for the portion of the globe between the tropics, this amounts to a yearly fall of 88,737 cubic miles, and that we can scarcely reckon the share of this great volume of water supplied by evaporation from the same part of the globe at more than one-half. Still less is it possible to calculate the amount of vapour annually transferred from the northern to the southern hemisphere, which goes to neutralize the apparent effect of the diversion of portions of the equatorial waters to the north side of the line. In the Atlantic basin it is probable that the larger part of the rainfall in the region including and surrounding the Gulf of Mexico and the Caribbean Sea is supplied by vapour carried from the temperate zone by the north-east trade-winds. There is some reason to believe that a portion of the rainfall of the great basin of the Amazons, south of the line, is also supplied from the same source. Several travellers report that during the rainy season the prevailing winds are from the west and north-west, the latter being especially predominant at

Iquitos, about 4° S. latitude, and 1600 miles from the mouth of the river.

In tropical Australia the rainy season falls during the prevalence of the north-west monsoon, and we cannot doubt that this is mainly supplied by vapour carried from the northern hemisphere. Another region wherein the same phenomenon is exhibited on a large scale is the central portion of Polynesia, extending from the Feejee to the Society Islands over a space of at least twenty degrees of longitude. Over that wide area, as far as about twenty degrees south of the line, the regular south-east trade-wind prevails only in the winter of the southern hemisphere, while during the rest of the year, especially in summer, north and north-east winds have the predominance. Taking the mean of three stations in the Feejee Islands, of which the returns are given by Dr. Hann, I find in round numbers the very large amount of 150 inches for the mean annual rainfall, of which 105 fall during the seven months from October to April, while the five colder months from May to September supply only forty-five inches of rain. There can be little doubt that the larger part of the 105 inches falling during the warm season is derived from the northern hemisphere.

I by no means seek to account fully for the apparent contradiction between the results of theory, as developed by Dr. Croll, and the actual distribution of heat over the earth as proved by observation; but I venture to think that I have shown reason to doubt the possibility of drawing absolute conclusions as to the results of astronomical changes until we shall have fuller knowledge than we now possess of all the agencies that regulate climates.

Before concluding these remarks, I will notice one other branch of the argument in regard to which I am unable to concur with Mr. Croll. As we have seen, the essential point in his theory as to the *modus operandi* of changes of eccentricity, and the relative position of the poles, on the distribution of temperature, is that the currents of the equatorial zone are driven towards the pole which has the summer in aphelion, and that the cause of this shifting of the currents depends on the greater strength of the trade-winds in the hemisphere which has the winter in aphelion; the strength of the trade-winds in turn depending on the amount of difference of temperature between the equatorial and the colder zones. Taking the surface of the earth generally, the trade-winds of the southern are probably stronger than those of the northern hemisphere, and, if it were true that the south temperate and frigid zones were colder than those of the other hemisphere, it would be allowable to argue that the greater difference of temperature as compared with the equatorial zone was the cause of the greater strength of the trade-winds. But we now certainly know that the southern hemisphere between latitudes 45° and 55° is considerably warmer than the corresponding zone of the northern hemisphere, and we have good grounds for believing that the mean temperature of the whole hemisphere south of latitude 45° is higher, and certainly not lower, than that of the same portion of the northern hemisphere. We are therefore not justified in explaining the greater strength of the southern trade-winds by a greater inequality of temperature between the equator and the pole.

In my opinion the cause of this predominance of the southern trade-winds is to be sought in the fact that the southern is mainly a water hemisphere, while the northern is in great part a land hemisphere. In the south, the great currents of the atmosphere flow with scarcely any interruption, except that caused by Australia, where, in fact, the trade-winds are irregular, and lose their force. In the northern hemisphere the various winds originating in the unequal heating of the land surface interfere with the normal force of the trade-winds, and weaken their effect.

In connection with this branch of the subject, I may remark that the belief in the greater cold of the southern hemisphere mainly rests on the fact that all the land hitherto seen in high latitudes has been mountainous, and is covered by great accumulations of snow and ice. But this does not in itself justify the conclusion that the mean temperature is extremely low. It is true that the fogs which ordinarily rest on a snow-covered surface much diminish the effect of solar radiation during the summer in high latitudes, but this is compensated by the great amount of heat liberated in the condensation of vapour. The only part of the earth which is now believed to be covered with an ice-sheet is Greenland, but the mean of the observations in that country shows a temperature higher by at least 10° Fahr. than that of Northern Asia, where the amount of snowfall is very slight, and rapidly disappears during the short arctic summer. If there be, as some persons believe, a large tract of continental land surrounding the south pole, I should expect to find that the great accumulations of snow and ice are confined to the coast regions. In that case the mean temperature of the region within the antarctic circle would probably be lower than it would be in the supposition, which appears to me more probable, that the lands hitherto seen belong to scattered mountainous islands. If, from any combination of causes, one pole of the earth has ever been brought to a mean temperature much lower than that now experienced, I should expect to find that the phenomena of glaciation would be exhibited towards the equatorial limit of the cold zone, rather than in the portions near the pole. The formation of land-ice depends on the condensation of vapour, and before air-currents could reach the centre of an area of extreme cold the contained vapour would have been condensed. This consideration alone suffices, to my mind, to make the supposition of a polar ice-cap in the highest degree improbable.

Mr. Wallace ("Island Life," p. 142) cites, as conclusive evidence of the effect of winter in aphelion in producing glaciation, the facts, to which attention was first directed by Darwin, as to the depression of the line of perpetual snow, and the consequent extension of great glaciers, on the west coast of Southern Chili. I have adverted to this subject in the text (p. 229), and I may further remark that if winter in aphelion be the cause of the depression of the snow-line in latitude 41° S., it can scarcely fail to produce some similar effect in latitude 34° S. Yet we find on the southern limit the snow-line much lower, and at the northern much higher, than it has ever been observed in corresponding latitudes in the northern hemisphere, the line being depressed by more than 8000 feet within a distance of only seven degrees of latitude. The explanation, as I have ventured to maintain, is altogether to be found in the extraordinary rainfall of Southern Chili; and to the same cause we must attribute the fact that, in spite of the greater distance of the sun, the winter temperature is higher than in most places in corresponding latitudes in the northern hemisphere. At Ancud in Chiloe, in latitude 41° 46', the temperature of the coldest month is lower by less than three and a half degrees of Fahrenheit than it is at Coimbra in Portugal, one and a half degree nearer the equator, in the region which receives the full warming effect of the Gulf-stream.

I should have expressed myself ill in the preceding pages if I should be supposed to deny that, in his writings on this subject, Mr. Croll has made an important contribution to the physics of geology. He has, in my humble opinion, been the first to recognize the full importance of one of the agencies which, under possible conditions, may have profoundly affected the climate of the globe during past epochs, although I do not believe that, in the present state of our knowledge, we can safely draw those positive inferences at which he has arrived. Even those who are unable to accept any portion of his theory as to the causes of past changes of climate must feel indebted to his writings for numerous valuable suggestions, and for the removal of many popular opinions which his acute criticism has shown to be untenable.

INDEX.

A

Acacia Cavenia, 157
Aconcagua, 192
—— valley, vegetation of, 195
Adesmia, 182
Agassiz, Professor Alexander, 342
Ajulla, Promontory of, 51
Albatross, 215
Alligators, 41
Alpine zone in Andes, 91
Amancais, 71
Amatapi, Sierra, 52
Ancud, 145
Andean Flora, Alpine zone, 104
——, divisions of the, 104
——, European genera common to, 101
Andean railways, 63
Andes, 49
——, Alpine zone in, 91
——, cactoid plant in, 92
——, Chilian, view of, 183
——, climate of Peruvian, 99
——, *Compositæ* in, 102
——, cosmopolitan weeds in, 101
Aneroid barometers, 353, 354
Angol, 213
Antarctic beech, 256
—— Flora, range of, 219
Anthopterus Wardii, 34
Anticosti, 273
Apoquinto, baths of, 188
Araucanian Indians, 212
——, language of, 213
Araucaria Brasiliensis, 311

Argentaria, climate of, 300
——, emigration to, 298, 299
——, forests of, 295
——, progress of agriculture, 298
——, frontier of Chili and, 258
Arica, vegetation of, 121
Armeria maritima, var. *andina*, 263
Artichoke, wild, 168
Atacama, desert of, 124, 131
Atlantic, colour of, 7
——, summer temperature of, 362
——, temperature of, 5
——, winds of, 345
Ayacucho steamship, 118
Azores, 5

B

Baccharis, 157
Bahia Blanca, 298, 357, 358
Bahia de Todos Santos, 346
Baillonia spartioides, 202
Balmacedo, Don F., 191
Banda Oriental, 281
——, vegetation of, 293
Barbadoes, absence of venomous snakes in, 14
——, black population of, 11
—— harbour police, 9
—— planters, 13
——, productiveness of, 8
Barometer, high, 4
——, tables for, 4
Beagle Channel, 273
Belem, Tower of, 363

Berberis buxifolia, 263
—— *empetrifolia*, 263
—— *ilicifolia*, 263
Berberry, 225
Bentos, Fray, 287
Bio-Bio river, 213
Black-fish, 7
Blue Mountains, Jamaica, 17
Bombax pubescens, 328
Borya Bay, 241
Bossi, Signor Bartolomeo, 281
Botafogo, 322
Bove, Lieutenant, 252
Bramble in Chili, 150
Brazil, ancient mountains of, 317
——, coffee-planting in, 341
——, geology of, 313, 314
——, glacial deposits in, 342
——, rainfall in coast region of, 334
Brazilian physicians, their fees, 340
Bridges, suspension, in the Andes, 85
Buenaventura, 32
Buenos Ayres, 293-295, 299

C

Cabo Blanco, 43
—— San Lorenzo, 37
—— Santa Elena, 38
Cachapoal river, 178
Cactoid plant in Andes, 92
Caldera, 133
Callao, 61
——, quarantine at, 57
Canary Islands, 362
Cape Froward, 243
—— Parinas, 44
—— pigeon (*Daption capensis*), 214
—— Pillar, 238
—— Verde Islands, 359
Capricorn, Tropic of, 132
Cardoon, 164
Casapalta, 91
Catamarans, 352
Cathartes atratus, 112
Caudivilla, 109
Cauquenes, Morro de, 182
——, town of, 169
Cauquenes Baths, 172
——, railway to, 163

Celery, wild, 221
Cereus Quisco, 151, 176
Cerro de Pasco, 72
—— del Roble, 151
Chacao, Canal de, 216
Chagres river, 22
Chañaral, 133
Channels of Patagonia, 222, 223
Chicla, hotel at, 80
——, scenery at, 84
——, vegetation of, 98
Chili and Argentaria, frontier of, 258
—— and Peru, naval war of, 58
——, bramble in, 150
——, Central, flora of, 141
—— ——, climate of, 143-145
—— ——, rainfall in, 144
——, European plants in, 164
——, physical geography of, 170
——, Southern, glaciers of, 229
—— ——, rainfall of, 229
Chilian elections, cumulative vote, 191
—— mines, 161
Chiliotrichium amelloides, 234
Chiloe, island of, 215
Chimborazo, 38
Chonos Archipelago, 216
Chosica, 73
Chuquiraga spinosa, 91
Churches in Lima, 62
Chusquea, 152
Cigars, Guayaquil, 41
Cinnamon tree, 10
Claraz, M. Georges, 357, 358
Clarence Island, 243
Climate, effects of tropical, 39
Cnicus lanceolatus, 164
Cobeja, 131
Coffee-planting in Brazil, 341
Colletia spinosa, 177
Colomba, 214
Colon, 21, 22, 27
Commercial travellers, German, 309
Compositæ in Andes, 102
Concepcion del Uruguay, 288
Condor, 87, 93
Condors, captive, 185
Copiapó, 133
——, Rio de, 133

INDEX. 409

Coquimbo, vegetation of, 136, 138
Corbett, Mr., 328
Cordillera de la Costa, 218
Cordillera Grande, of Goyaz, 315
Cordillera Pelada, 218, 219
Cordillera in Peru, 49
Corrientes, 294, 300
Cosmos Line, German steamers of, 205
Cousiño, Madame, 207
Crab, red, 227
Croll, Dr. James, 271
——, remarks on his theory of secular changes of climate, 393
Cryptocarya Peumus, 160
Cuyabà, 279, 310

D

Dandelion (*Taraxacum lævigatum*), 263
Daption capensis, 214
Darwin, Mount, 267, 270
Dawson Island, 245
Desfontainea spinosa, 225
Desolation, Land of, 235, 241
Diomedea exulans, 215
—— *fuliginosa*, 215
Don, Royal Mail steamer, 2
Doterel, wreck of the, 267
Drimys Winteri, 147
Drummond-Hay, Mr., 148
Dungeness, 271
Duvaua dependens, 293

E

Earthquake-waves, 122
Eccremocarpus scaber, 199
Ecuador, 36, 40
Eden harbour, 224
Education, Chilian zeal for, 265
Elections, Chilian, cumulative vote, 191
Encelia canescens, 134
Engler, Dr., 34, 106
English Narrows, 223
English the *lingua franca* of America, 88
Ensenada, 302
Entrerios, 288

Equator, cold current near, 356
——, path of the sun, 37
Equatorial rains, 352
—— vegetation, 33
Erodium cicutarium, 165
Escallonia, 181
Espiritu Santo, Cape, 271
Eucalyptus globulus, 160, 292
Evergreen beech (*Fagus betuloides*), 225
Existence, struggle for, 330
Eyre Sound, 228

F

Fagus betuloides, 225
—— *obliqua*, 151
Falkland Islands, 247, 273
Fayrer, Sir Joseph, 349
Fenton, Dr., 250, 262, 269
Fernando Noronha, 354, 355
Feuillée, Father, 184
Flint, Mr., 153
Flowering plants, origin of, 318
Flying-fish, 5
—— of Pacific, 53
Fogs on Peruvian coast, 54
Francoa sonchifolia, 210
French, Dr., 289, 291
Fruit-sellers, migratory, 42
Fuegians, 233, 242, 260, 261

G

Gallinazo, 87
——, scavenger bird, 112
Galvesia limensis, 45
Gillies, Captain, 344
Glacial deposits in Brazil, 342
Glaciers in South Patagonia, 239
—— of Southern Chili, 229
Glaziou, Dr., 324
Gleichenia, 226, 311
Glyptodon, 358
Gongo Seco, 316
Gordontown, Jamaica, 18
——, cool climate of, 19
Graham, Mr. J. R., 67
Granite, disintegration of, 315
Grisebach, 34, 145
Gualtro, 168

Guanacos, 131, 253
Guano Islands, 53
Guayaquil cigars, 41
Guayaquil, city of, 40, 41
——, Gulf of, 38, 40
Guayas river, 38, 41, 42
Gynopleura linearifolia, 157

H

Hale Cove, 220
Hann, Dr. Julius, 144, 305, 349
Hanover Island, 236
Hayti, island of, 15
——, cannibalism in, 16
Haze, opacity of, 158
Heights above sea-level, fall of temperature in ascending to, 369
Henderson's Inlet, 231
Hess, Mr., 163
Hippeastrum equestre, 19
Hopedale, in Labrador, 273
Huanillos, 127
Humboldt current, 50
Humming-birds, 209
Hura crepitans, 10
Hymenophylla, 226

I

Iberia steamship, 269
Ice-axe, 226
Ice, floating, 228
Illiluk, 274
Immigrant plants, Darwin's view of, 166
——, checks on their extension, 167
Indians, Araucanian, 212
——, Patagonian, 260
Iquique, 121
——, sea-fight at, 127
Isla de Santa Maria, 211
Islay steamship, 29
Itamariti, Falls of, 329

J

Jamaica, 16
——, black population of, 20
——, vegetation of, 18
Jacmel harbour, 15

K

Kageneckia oblonga, 175
Kingston, Jamaica, 16

L

Lapageria rosea, 209
La Plata, estuary of, 277, 283
Las Condes, 163
La Serena, 136, 145
Lavapie Promontory, 211
Liais, M., 342
Libocedrus, 235
Lichens, 221
Lima, 61
——, a dinner-party at, 108
——, ancient beaches near, 113
——, meteorological observations at, 99
Liriodendron, 344
Lisbon, Rock of, 363
Llaillai, 150
Llama in Peru, 95
Loa river, 127
Lobelia gigantea, 184
Lobelia tupa, 184
——, poisonous species of, 76
Lobos de tierra, 53
—— de afuera, 53
Lomaria magellanica, 226
Lombardi, Signor M., 109
Lombardy poplar, 160
Loranthus, 176, 202
Lord Nelson Strait, 236
Lota, coal deposits of, 207
——, parque of, 208
Lynch, Don Patricio, 67
——, his administration, 68

M

Maceio, 347
Magellan, Straits of, 238, 239, 367
——, forests in the, 240
——, variable climate in, 240, 254
Maipo river, 134
Maldonado, 304
Malesherbiaceæ, 157
Mango tree, 10
Mapocho river, 157

Markham, Captain Albert, 135
Marrubium vulgare, 138
Matto Grosso, 279
Matucana, San Juan de, 76
Mayne Channel, 237
Maytenus magellanica, 263
Meiggs, Mr., 65
Mejillones, 132
Memory, lapses of, 26
Mendoza, 300
Mercator's projection, 30
Messier's Channel, 220, 231
Mist, clearing of, 333
Mitraria coccinea, 225
Molina, 175
Mollendo, 119
——, a bad port, 120
Monkeys, domesticated, 332
Monson, Mr., 280
Montaña of Eastern Peru, 97
Monte Video, 277, 279
Morro de Cauquenes, 182
Mountain-sickness, 81
Mulinum, 189
Mutisia, 177
Mutisiaceæ, 102
Myzodendron punctulatum, 256

N

Napp, Mr. Richard, 300
Nation, Mr. W., 70, 108
Naval war of Chili and Peru, 58
New Granada, 32
Nicotiana glauca, 360
Nikolaiewsk, 273
North Atlantic, trade wind of, 361
Northern hemisphere, temperature of, 273, 274

O

O'Higgins, General, 154
Olfactory nerve, fugitive impressions, 190
Oreodoxa regia, 324
Organ Mountains, 325
Oroya railway, 64
——, spiral tunnel of, 79
——, viaducts of, 74
Ostrich, South American, 261
Oxalis lobata, 146

P

Pacific coast-steamers, 31
Pacific, colour of water of, 31
——, first view of, 25
——, flying-fish of, 53
——, high seas in Southern, 211
—— steamer, delay of, 269
Paisandu, 288, 289
Palms, avenue of, 323
Panama, 21
—— Bay, birds in, 29
—— Grand Hotel, 27
—— railway, 25
—— ship-canal, 23
——, vegetation of, 24
Paraguay river-steamers, 279
Paraná, 317
——, basin of the, 312
—— river, 284
Paranagua, Bay of, 308
Paranahyba, 313
—— valley, 320
Parasites and climbers, 330
Patagonia, 300
—— Channels, scenery of, 222, 227
——, vegetation of, 225, 235
——, women of, 253
Patagonian coast, winter climate, 276
Patagonian Indians, 260
Payta, climate and vegetation of, 45
Peckett Harbour, 262, 270
Pedro, Dom, Emperor of Brazil, 337
Pelicans, black, 59
Peñas, Gulf of, 218
Pernambuco, 351
Pernettya, 225
Peru, 44
—— and Chili, naval war of, 58
——, climate of Northern, 47
——, future of, 117
Peruvian coast, fogs on, 54
——, low temperature of, 55
—— sugar-plantation, 110
Pessimism, 365, 366
Petrel, giant, 215
Petropolis, 326, 327, 335
——, hermit of, 331
——, winter climate of, 334

INDEX.

Peumo tree, 159, 175
Philippi, Dr., 154
———, Professor Federigo, 155, 219
Physicians, Brazilian, their fees, 340
Pierola, Dictator of Peru, 117
Pietrabona, Commander, 257
Pimento tree, 10
Pisagua, 123
———, white rocks at, 125
Pisco, 119
Plantago maritima, 263
Pleroma arboreum, 342
——— *granulosum*, 336
Podocarpus nubigena, 235
Poncho, 179
Porlicra hygrometrica, 199
Port Famine, 245, 250
——— Gallant, 241
Potato, wild, in Andes, 93
Prado, General, 36
Prosopis limensis, 46
Proustia Baccharoides, 195
Pteris aquilina, 329
Puente Infernillo, 78
Puerto Bueno, 233
Punta Arenas, 145, 246, 273
Puya, 151

Q

Quadras in Santiago, 153
Quarantine at Callao, 57
——— at St. Vincent, 359
Quaresma, 336
Queen Adelaide Island, 236, 238
Quillaja saponaria, 175
Quillota, Valley of, 150
Quinta Normal at Santiago, 155

R

Railways, Andean, 63
———, Oroya, 64
———, spiral tunnel of, 79
———, viaducts of, 74
Rancagua, 168
Reed, Mr. Edwin, 204
Reilly, Mr., 210
Resguardo del Rio Colorado, 200
Rhamses, the, 205

Rhea Darwinii, habits of, 261
Rimac, valley of the, 71
———, ancient terraces in, 75
———, *Compositæ* in, 76
———, effects of sea-breeze in, 98
Rio Claro, 171
——— Colorado, 276
——— Janeiro, Bay of, 321, 322, 325
——— Parahyba do Sul, 313
——— San Francisco, 314
Rocks, disintegration of, 115
———, ice-action on, 228
Rumex acetosella, 263

S

Sagittaria Montevidensis, 297
Saladeros, 287
Salix Humboldtiana, 77
Salta, 294
Salto, 291
Sambucus Peruviana, 101
Sampayo, Don Francisco, 257
San Bartolomé, 73
——— Cristobal, Cerro, 156
Sand-box tree (*Hura crepitans*), 10
Sandy Point, 246, 250
———, burnt forest at, 256
———, mutiny of convicts at, 251
———, the hotel at, 249
———, vegetation of, 255, 263
San Felipe, 192
Sanitary rules, neglect of, 87
San José, Promontory of, 276
——— volcano, 164
San Lorenzo, island of, 59
San Matias, Bay of, 276
San Paulo, 308–310
——— and Rio Janeiro railway, 312
———, railway from Santos to, 307, 308
San Ramon, Salto de, 190
Santa Clara, 72
Santa Cruz settlement, 246
Santa Lucia, Rock of, 162
Santa Rosa de los Andes, 193, 196
Santiago, 145, 153, 156, 161
———, railway to, 149
———, sunset at, 186

INDEX. 413

Santos, 305, 306
——, tropical vegetation at, 307
São João da Barra, 312
São Salvador, 346
Sarmiento Channel, 231
Sarmiento, Mount, 243, 244, 267, 270
Scavenger bird, 112
Schinus molle, 77
Sea-sickness, 217
Seaweed, bands of, 6
Senecio, the genus, 268
Serra da Mantiqueira, 313, 314
Serra do Mar, 308, 314
Shannon, Captain, 269
Simpson, Captain, 207
Sitka, 274
Smyth's Channel, 231, 237
Solanum mammosum, 33
Soroche, mountain-sickness, 81
Soto, Don Olegario, 171
South America, tropical, origin of flora, 35
——, rainless zone of, 48
South Brazil, plateau of, flora, 311
South Patagonia, glaciers in, 239
Southern Atlantic, climate of, 305
Southern Cross, 7, 253
Southern hemisphere, temperature of, 272–274
Spanish-Americans, indolence of, 89
Species, groups of incomplete, 181
Staten Island, 252, 258
St. Antao Island, 359
Steamers, Pacific coast, 31
Straits of Magellan, 270
Sunstroke, causes of, 349, 350
Surco station, 75
Swinburne, Don Carlos, 154

T

Taforò, Dr., 191
Tagus steamship, 344
Talca, 145
Taltal, 133
Tamar, Cape, 240
Tambo de Mora, 119
Tarapacà, 125
Taraxacum lævigatum, 263

Telephone, use of, in South America, 290
Tierra del Fuego, 245, 267
Tijuca, 338
——, giant tree near, 343
——, vegetation of, 341
Tillandsia, 307
Titicaca, Lake of, 63, 66
Tocantins river, 315
Tocopilla, 128, 133
——, scenery of the moon, 129
Trade wind, north-east, 6
Trescott, Mr., 60
Tres Montes, Cape, 218
Trinidad, Gulf of, 231
Triumph, the ship, 135
Tropæolum tuberosum, 78
Trumpet-flower (*Bignonia venusta*), 307, 311
Tucuman, 294
Tumaco, 36
Tumbez, 43
Tupa Berterii, 184
—— *secunda*, 184
Tupungato, the Peak of, 153

U

Ucayali river, English settler at, 96
Unalaschka, 274
Uruguay, climate of, 279
——, fossil remains in, 291
——, islands of the, 287
—— Republic of, chronic disorder, 282, 284, 285
Ushuaia, mission station at, 260
Ushuaja, 273
Uspallata Pass, 200
Utricularia, 33

V

Valdivia, 145
Valparaiso, 138, 145
——, danger of earthquakes at, 139
Vegetation, equatorial, 33
Verbena family, 201
Viaducts, Oroya railway, 74

Vicuña Mackenna, Don Benjamin, 158
Villages, remains of ancient Peruvian, 73
Viña del Mar, 149
Vincent, St., aspect of, 360
——, quarantine at, 359
Vinciguerra, Signor, 252
Virgenes, Cape, 271
Volcano de Chana, 219

W

Wellington Island, 222
Willsen, Captain, 205
Winter's bark (*Drimys Winteri*), 147

Y

Yellow fever, treatment of, 339

NOTE ON THE MAP OF SOUTH AMERICA.

IN the annexed map an attempt has been made to represent the probable course of the isothermal lines—lines denoting equal temperature—in the South American continent. The black lines indicate the mean temperature for the entire year; the red lines that for January, the hottest month; and the green lines that of July, the coldest month. The numbers placed over each line in corresponding colours indicate the temperature in degrees of the Centigrade scale. We possess a fair amount of information as to the meteorology of the coasts of the continent; but of the interior our knowledge is miserably deficient, and is nearly limited to several stations in Argentaria, and a few in the basin of the Amazons. As a result, the course of the isothermal lines in the interior is to a great extent conjectural. As in all similar maps, no account has been taken of the relief of the surface; when a line crosses a mountain range, the temperature indicated is that which would be found, as is assumed, if the height were reduced to the sea-level. No attempt has been made to show the variations of temperature with the season in the part of the continent near the equator. These are very slight, and depend mainly on local conditions, the mean temperature of the year varying from $25·5°$ to $28°$ C., or from about $78°$ to $82°$ Fahr.; the hottest seasons near the equator, apart from local conditions, being those of the equinoxes.

The chief interest of the map to the physical geographer arises from the remarkable effect of the southern, or Humboldt, current, in lowering the temperature of the western coast between the fifth and the fortieth degrees of south latitude. This is, of course, most apparent in the isothermal for January. It will be seen that at that season the temperature of Northern Peru is about the same as that of Buenos Ayres, lying thirty degrees farther from the equator. In midwinter (July) the effect is far less apparent, and in the south of the continent the isotherms for that season nearly correspond with the parallels of latitude. The lines indicating mean annual temperature naturally assume a course intermediate between those for the extreme seasons.

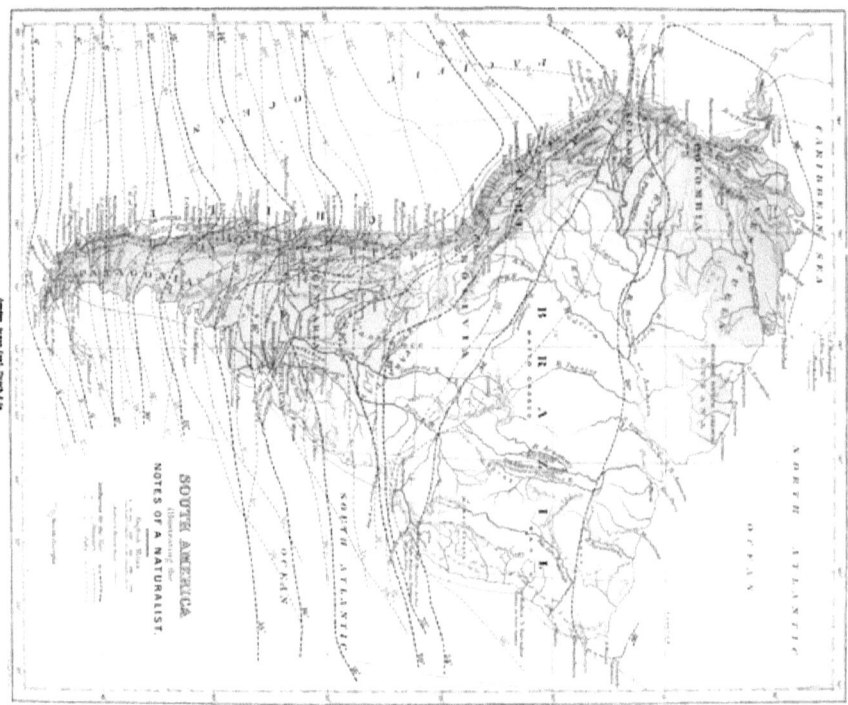

www.ingramcontent.com/pod-product-compliance
Lightning Source LLC
Chambersburg PA
CBHW051736300426
44115CB00007B/594